Planning the Oregon Way

Planning the Oregon Way

A Twenty-Year Evaluation

edited by

Carl Abbott
Deborah Howe
Sy Adler

Oregon State University Press
Corvallis, Oregon

The paper in this book meets the guidelines for permanence and durability of the Committee on Production Guidelines for Book Longevity of the Council on Library Resources and the minimum requirements of the American National Standard for Permanence of Paper for Printed Library Materials Z39.48-1984.

Library of Congress Cataloging-in-Publication Data

Planning the Oregon way : a twenty-year evaluation / edited by Carl Abbott, Deborah Howe, Sy Adler

    p. cm.

    Includes bibliographical references and index.

    ISBN 0-87071-381-7

    1. Land use—Government policy—Oregon. I. Abbott, Carl. II. Howe, Deborah A., 1953- . III. Adler, Sy, 1950- .
HD211.O7P55 1993
333.73'13'09795—dc20　　　　　　　　　93-36408
　　　　　　　　　　　　　　CIP

Copyright © 1994 Oregon State University Press
All rights reserved
Printed in the United States of America

Contents

Introduction ix

Part I
Building the Oregon System

CHAPTER 1
Land Use Politics in Oregon **3**
Gerrit Knaap

CHAPTER 2
Oregon's Urban Growth Boundary Policy as a Landmark Planning Tool **25**
Arthur C. Nelson

CHAPTER 3
The Legal Evolution of the Oregon Planning System **49**
Edward J. Sullivan

CHAPTER 4
Irreconcilable Differences: Economic Development and Land Use Planning in Oregon **71**
Matthew Slavin

Part II
Planning Issues and Choices

CHAPTER 5
Housing as a State Planning Goal **91**
Nohad A. Toulan

CHAPTER 6
The Oregon Approach to Integrating Transportation and Land Use Planning **121**
Sy Adler

CHAPTER 7
Siting Regional Public Facilities **147**
Mitch Rohse & Peter Watt

CHAPTER 8
Oregon Rural Land Use: Policy and Practices **163**
James R. Pease

CHAPTER 9
Land Use Planning and the Future of Oregon's Timber Towns **189**
Michael Hibbard

Part III
Perspectives and Interpretations

CHAPTER 10
The Oregon Planning Style **205**
Carl Abbott

CHAPTER 11
Following in Oregon's Footsteps: The Impact of Oregon's Planning Program on Other States **227**
John M. DeGrove

CHAPTER 12
Managing "the Land Between": A Rural Development Paradigm **245**
Robert C. Einsweiler & Deborah A. Howe

CHAPTER 13
A Research Agenda for Oregon Planning: Problems and Practice for the 1990s **275**
Deborah Howe

Afterword **291**

Oregon's Statewide Planning Goals **299**

Annotated Bibliography **305**

Contributors **319**

Index **323**

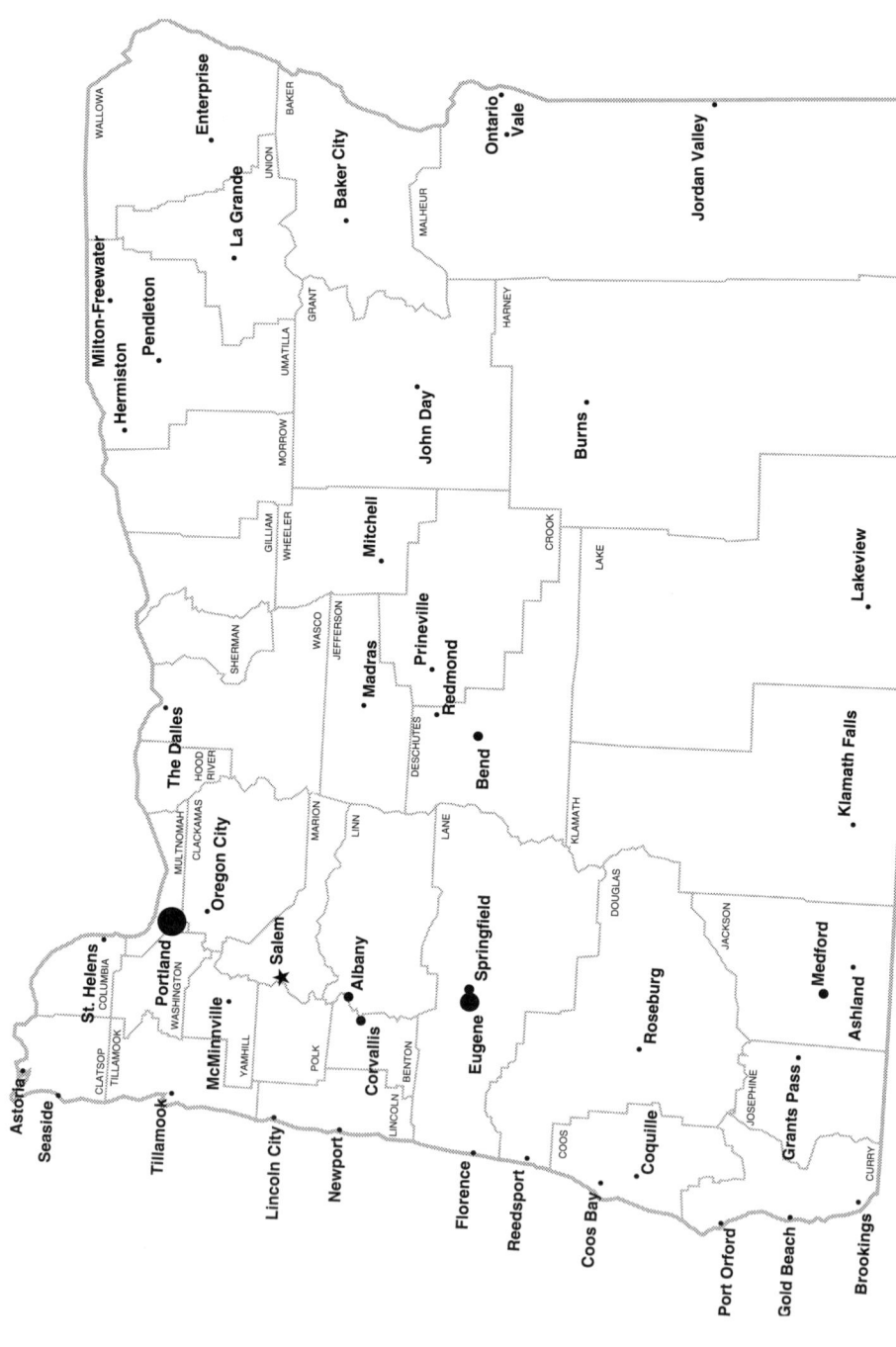

Base map of Oregon counties courtesy of A. Jon Kimerling

Introduction

In 1973, Oregon took a pioneering step in land use planning. Signed into law on May 29, 1973, Oregon Senate Bill 100 created an institutional structure for statewide planning. It required that every Oregon city and county prepare a comprehensive plan in accordance with a set of general state goals. While preserving the dearly held principle of local responsibility for land use decisions, it simultaneously established and defined a broader public interest at the state level. Supervised by a Land Conservation and Development Commission (LCDC), the Oregon system has been an effort to combine the best of these two approaches to land use planning. The very existence of Oregon's planning system has helped to inspire and justify similar programs elsewhere. Its details have been studied, copied, modified, and sometimes rejected as Florida, Maine, New Jersey, Georgia, and other states have considered "second generation" systems of state planning.

The twentieth anniversary of the Oregon system marks an opportune time for reflection and evaluation. To this end we have invited both academic experts and practitioners to comment on the Oregon experience. John DeGrove of Florida Atlantic University provides the perspective of someone who has studied statewide planning systems for over twenty years. Gerrit Knaap of the University of Illinois and Arthur C. Nelson of the Georgia Institute of Technology have studied Oregon's program extensively, while Robert Einsweiler of the Lincoln Institute of Land Policy brings a broad comparative perspective on regional growth issues. Several of the contributors have been involved in both teaching and consulting about Oregon planning issues as faculty members at the University of Oregon (Michael Hibbard), Oregon State University (James Pease), and Portland State University (Carl Abbott, Deborah Howe, Sy Adler, Nohad Toulan). Practitioners are represented by Ed Sullivan, recognized as the state's leading land use attorney, by Mitch Rohse of the Department of Land Conservation and Development, and by Peter Watt of the Lane Council of Governments.

The first section of the book covers the evolution of the planning system from the 1970s to the 1990s. Specific issues such as housing, transportation, public facility siting, rural lands, and economic restructuring are examined next. The third section explores the future of the Oregon system, its relevance to other states, and directions for change. Although the editors have worked to eliminate unnecessary overlap among the chapters, we have not tried to reconcile conflicting interpretations of the Oregon system. Indeed, we believe that one of the values of this collection is the presentation of multiple points of view.

There tend to be two mindsets among those who are favorably disposed toward the Oregon planning system. Outsiders frequently think it is extraordinary, in part because they have the perspective of trying to plan in environments that do not value planning or do not provide the institutional context that facilitates coordination, collaboration, and continuity over time. Insiders are in a position to see the flaws. They are mired in minutiae and are painfully aware of the program's inadequacies. They may no longer have the perspective of what it is like to work in a system that does not have a broader framework. Indeed, a whole generation of Oregon planners has experience with only this system. Many planners are frustrated with state rule making, the role of the Land Use Board of Appeals in interpreting requirements, and other detailed legal processes so clearly described by Ed Sullivan. In their minds, these technicalities put proactive planning on the back burner.

It is valuable to note as context that a legalistic orientation to planning and regulation reflects the strength of the environmental protection movement. As Vogel (1985) has argued more generally for the United States, the environmental movement has typically sought highly detailed rules and has leaned heavily on the courts to counterbalance the perceived power of development interests at the local level. In the case of Oregon land use planning, the advocacy group 1000 Friends of Oregon has continually pressed for vigilant enforcement of strong statewide regulations.

As Gerrit Knaap notes, however, the program provides the framework for the ongoing resolution of challenges. Carl Abbott's ideas regarding the culture of planning suggest that underlying support for Oregon's approach to public policy making is strong and likely to continue. It is certainly true that strong leadership for the Land Conservation and Development Commission itself has been drawn from all parts of the state. In addition, the state has shown flexibility by adopting an increasingly fine-grained approach in its interventions, mandating different categories of actions in different areas, and even excusing some places from compliance. Sy Adler's chapter on the transportation rule describes one of the ways in which the system is being modified to respond to specific issues. James Pease's chapter on rural lands discusses the effort to develop regulations that can be adapted to different local circumstances.

This willingness by the state to fine-tune planning requirements reflects political circumstances at the local level. Portland, other Willamette Valley cities, and other large jurisdictions have the technical and political capacity to address a wide range of planning issues. They are able to differentiate themselves and to make arguments about the varying relevance of state mandates to local circumstances. As shown by the debate over less productive resource lands, small jurisdictions with limited resources of time and staff expertise may have had a more difficult time in articulating their cases and justifying flexible responses.

In presenting a wide range of ideas on the Oregon planning system, we hope to facilitate a debate and synthesis between the perspectives of outsiders and insiders. Outsiders need to have a more realistic understanding of the challenges that Oregon is facing and the mechanisms that are emerging to address the challenges. Insiders need affirmation of the program's potential and progress in adapting it to new circumstances. If Oregonians can see the context within which they work then they can have a better sense of why certain changes are needed and how these can be accomplished.

Getting to the Goals

When the legislature adopted Senate Bill 100, formal land use planning in Oregon was just over fifty years old. The state's initial planning legislation in 1919 and 1923 granted cities the authority to develop plans and land use regulations. In a 1920 referendum, Portland voters narrowly rejected citywide zoning under the first enabling act. Four years later, they overwhelmingly approved a simpler zoning ordinance. Planning remained solely a city function until 1947, when the legislature extended similar authority to counties in response to chaotic growth of urban fringe areas during the boom years of World War II. Counties were authorized to form planning commissions, which could recommend "development patterns" (renamed "comprehensive plans" after 1963). Counties, unlike cities, were required to develop zoning and other regulations to carry out their plans. The concern with disorderly growth that led to county planning in the 1940s grew into serious worries about suburban sprawl as Oregon began to grow rapidly in the 1960s. By the end of that decade, Willamette Valley residents from Eugene to Portland viewed sprawl much more broadly as an environmental disaster that wasted irreplaceable scenery, farm land, timber, and energy. Metropolitan growth was explicitly associated with the painful example of southern California. Governor Tom McCall summarized the fears of many of his constituents in January 1973, when he spoke to the Oregon legislature about the "shameless threat to our environment and to the whole quality of life—unfettered despoiling of the land" and pointed his finger at suburbanization and second home development.

McCall had already presided over six years of environmental protection. Behind his dramatic flair was a sincere and long-term concern about pollution and sprawl. "Pollution in Paradise," a television documentary about the Willamette River that McCall filmed in 1961-62, made a lasting impression on McCall himself as well as its TV audience. During McCall's first term as governor (1967-70), he created a state Department of Environmental Quality, started planning for a Willamette River Greenway, and presided over passage of bills to reassert public ownership of ocean beaches, to set minimum deposits for beverage cans and bottles, and to require removal of billboards.

In this context of environmental awareness, the initial impulse for state land-use legislation came from the farms rather than the cities.[1] The center of concern was the hundred-mile-long Willamette Valley, where the Coast Range on one side and the high Cascades on the other reminded residents that land is finite. The first steps toward the idea of "exclusive farm use" between 1961 and 1967 involved legislative action to set the tax rate on farm land by land rental values—in effect, by its productive capacity as farm land—rather than by comparative sales data which might reflect the demand for suburban development. A conference on "The Willamette Valley—What Is our Future in Land Use?" held early in 1967 spread awareness of urban pressures on Oregon's agricultural base. With key members drawn from the ranks of Oregon farmers, the Legislative Interim Committee on Agriculture responded by developing the proposal that became Senate Bill 10, Oregon's first mandatory planning legislation.

Adopted in 1969, SB 10 took the major step of requiring cities and counties to prepare comprehensive land-use plans and zoning ordinances that met ten broad goals. The deadline was December 31, 1971. However, the legislation failed to establish mechanisms or criteria for evaluating or coordinating local plans, allowing some counties to opt for pro forma compliance. In a 1970 referendum, 55 percent of the state's voters expressed support for SB 10. At the same time, McCall's successful re-election campaign called for strengthening the law.

When the leadership of the 1971 legislature blocked formation of a formal interim study committee, Senator Hector Macpherson, a Linn County dairy farmer, worked with McCall to set up an informal Land Use Policy Committee to suggest ways to improve SB10. Members of the committee represented the governor's office, environmental groups, and business organizations. The Oregon legislature acted in 1973 to correct flaws in the 1969 law. A state-sponsored report by San Francisco landscape architect Lawrence Halprin, "Willamette Valley: Choices for the Future," helped to set the stage in the fall of 1972. McCall's "grasping wastrels" speech with its anathema on unregulated land development raised the curtain. Greatest credit for passage of SB 100 went to Senator Macpherson, who was convinced of the need to fend off the suburbanization of the entire valley. Drawing on his experience on the Linn County Planning Commission, he articulated the importance of a statewide planning program in protecting and enhancing agricultural investment. This argument served to dampen the demands of farmers to preserve property rights that would enable them to sell out to developers.[2]

As Macpherson later recalled, "our bible when we were putting the thing together" was Fred Bosselman and David Callies's book, *The Quiet Revolution in Land Use Control*, published for the federal Council on Environmental Quality in 1972. The volume described state-level land planning programs in Hawaii and Vermont and a number of state efforts to protect such environmentally sensitive lands as Massachusetts wetlands and Wisconsin shorelands. Perhaps the central message for those crafting the Oregon legislation was the need for state programs to incorporate continuing local participation.

In the 1973 legislature, essential help came from Senator Ted Hallock of Portland, from Representative Nancie Fadeley, and from L. B. Day, a Teamster's Union official representing Willamette Valley cannery workers and a former director of the state Department of Environmental Quality. Hallock and Fadeley chaired the Senate and House committees on environment and land use. Day was the dominant influence among a task force of lobbyists whom Hallock called together to hammer out necessary compromises. Fierce opposition forced the deletion of two major provisions from the draft legislation. One was the designation of "areas of critical state concern" where the state would have overriding control. The other was the designation of councils of government rather than counties to coordinate local plans. The final version of SB100 passed the Senate by eighteen votes to ten. Fadeley's committee agreed to Macpherson's plea to report the bill to the House floor without changes, thus avoiding the minefield of a conference committee. In total, forty-nine out of sixty legislators from Willamette Valley districts voted in favor of SB 100. Only nine of their thirty colleagues from coastal and eastern counties did so.

Passage of the bill in May 1973 created the Land Conservation and Development Commission to oversee compliance of local planning with statewide goals. The commission is composed of seven members appointed for four-year terms by the governor and confirmed by the State Senate. One member is appointed from each of Oregon's five congressional districts and two from the state at large. At least one but no more than two members must be from Multnomah County, the state's largest and most urban county. At least one member must be an elected city or county official at the time of appointment. Beginning in 1997, the membership will have to include one elected county official and one current or former elected city official at the time of appointment. Staff support for LCDC and the planning program comes from the Department of Land Conservation and Development.[3]

As its first task, the new LCDC rewrote the state planning goals in 1974 after dozens of public workshops throughout the state. The ten goals of the 1969 legislation were made more clear and precise and four new goals were added. All fourteen goals were adopted by LCDC in December 1974. An additional goal on the Willamette River Greenway was added in December 1975 and four goals focusing on coastal zone issues were added in December 1976. The goals, often referenced by number rather than name, are as follows (see the Appendix for the full wording):

1. Citizen Involvement
2. Land Use Planning
3. Agricultural Land
4. Forest Lands
5. Open Spaces, Scenic and Historic Areas, and Natural Resources
6. Air, Water, and Land Resources Quality

7. Areas Subject to Natural Disaster and Hazards
8. Recreational Needs
9. Economy of the State
10. Housing
11. Public Facilities and Services
12. Transportation
13. Energy Conservation
14. Urbanization
15. Willamette River Greenway
16. Estuarine Resources
17. Coastal Shorelands
18. Beaches and Dunes
19. Ocean Resources

The basic idea behind the program is that development is to be concentrated within urban growth boundaries (UGBs) which are established around incorporated cities. Outside of the UGBs are resource lands where land use policies are aimed at supporting the vitality of the agricultural and forest industries. Development unrelated to resources is strictly limited in resource areas.

Oregon's land use program matured between 1974 and 1982. As Edward Sullivan's chapter describes, implementation required procedural innovations. LCDC defined a formal process of "acknowledgment" to certify that local plans actually met state goals. It similarly defined a requirement for "periodic review" to make sure that plans were adapted to changing circumstances. The legislature established the Land Use Board of Appeals (LUBA) as a specialized appellate court to deal with the increasingly complex details of land use law and cases. Local jurisdictions struggled to meet LCDC deadlines, adding staff to small or nonexistent planning offices. The first local plans were acknowledged in 1976, the last nearly a decade later.

The program also survived three initiative challenges, winning voter approval by a margin of 57 percent to 43 percent in 1976 and 61 percent to 39 percent in 1978. Support was strongest in Portland, Salem, and Eugene. In 1978, the LCDC program also gathered support along the northern coast and in south-central counties where rapid recreational development had brought problems of urban services. Editorial discussion throughout the state emphasized the issue of local control. A few newspapers such as the *Newport News-Lincoln County Times* (October 20,

1976) and the Klamath Falls *Herald and News* (October 22, 1976) believed that voters should return planning to the localities. The more common editorial position was that the LCDC process protected public participation and assured that local residents did the necessary land use planning. At the time of the 1978 referendum, support for LCDC could be found in newspapers serving the state's largest cities of Portland, Salem, Eugene, and Medford; in tourist-oriented cities of central Oregon such as Bend and Redmond; and in larger eastern Oregon communities such as Pendleton and Baker.[4]

During the depression of 1981-82, however, LCDC became the target of frequent complaints that planning requirements inhibited economic development. Opponents of the state planning system placed an anti-LCDC measure on the November 1982 ballot, calling for the abolition of LCDC, return of all land use planning authority to localities, and retention of state goals purely as guidelines. Editorial discussion now debated the economic impacts of statewide planning. Most newspapers agreed with the Bend *Bulletin* (October 17, 1982) that Measure 6 was irrational scapegoating. With some exceptions east of the Cascades, most editorial writers accepted the view of planning proponents that statewide planning actually encouraged economic development by requiring the designation of industrial land, stimulating tourism, and allowing large corporations to make plans for the long term. Although "opponents of the planning program use it as a scapegoat for Oregon's depressed economy," commented the Eugene *Register-Guard* (October 10, 1982), "those sincerely concerned with promoting economic development in Oregon should cheer this program rather than fight it."[5] A task force headed by Umatilla County farmer Stafford Hansell heard testimony from more than four hundred Oregonians and reported essentially the same conclusions to Governor Vic Atiyeh. The election returns showed the same regional divisions as before, with strong opposition from ranching counties in the southeastern corner of the state and from lumbering counties in the southwestern corner.

The 1982 referendum was the last comprehensive attack on the Oregon planning system. The rest of the decade brought institutional stability. A continuing economic slump triggered net outmigration that totaled 86,000 from 1980 to 1986. Stagnant population meant little demand for new housing and few pressures for land conversion, leaving the assumptions of most local plans unchallenged. Local planning activities focused on updates through periodic review rather than reexamination of basic goals. In addition, as Mitch Rohse and Peter Watt describe, the legislature tried to blunt potential opposition to the Oregon system by developing alternative procedures for deciding the location of controversial public facilities such as prisons.

A New Generation of Planning Issues

New challenges to the state planning system have come with the 1990s. The state attracted more than 100,000 in-migrants in the two years ending July 1991. Many of the newcomers have chosen metropolitan Portland, which anticipates substantial continued growth over the next two decades. Expansion of tourism and the popularity of Oregon for California retirees have also brought growth pressures to coastal and southern Oregon and the east slope of the Cascades. Meanwhile, passage of a property tax limitation measure in 1990 put a cap on local tax rates and transferred responsibility for a substantial portion of school funding to the state. This is threatening deep cuts in state and local services including land use planning. At the same time, the state legislature is putting its weight behind implementation of Oregon Benchmarks, a set of numerical objectives intended to serve as measures of the quality of life in Oregon. State agencies are being held accountable for achieving these objectives in a resource-poor environment. These demographic and economic trends along with a demanding yet constrained political climate underscore the need to take a new look at some of the issues that lie at the heart of the LCDC system. It has never been more imperative to demonstrate the effectiveness and efficiency of the land use planning process.

As Chris Nelson reveals, the urban growth boundaries are proving to be effective. Oregon has avoided the situation of a superficially comparable state like Colorado, where exurban and second home development has scarred vast sections of the Front Range. Despite Oregon's relative success, however, issues of intergovernmental coordination, facility planning, and planning for long-term UGB expansion remain unresolved.

In spite of the existence of a strong planning program, parts of Oregon are realizing low-density development patterns similar to those found elsewhere in the United States. The market continues to produce large single-family residences on the premise that they are what people want. Oregon policy makers and planners are actively exploring ways to intervene in this trend and to effect greater densities.

The transportation rule is an outgrowth of frustrations with the historical separation of land use and transportation planning. As Sy Adler indicates, the hoped-for integration of land use and transportation should achieve more purposeful urban form and a land use pattern that provides a range of mobility alternatives. Similarly, a gradual convergence of land use planning and economic development policy is occurring. Especially in major urban areas, as John DeGrove points out, large economic interests acknowledge the value of a stable planning environment for long-term investment. Economic development is increasingly a central element in neighborhood and district planning in the state's dominant city of Portland.

At the same time, the continued economic crisis of resource-dependent communities has created the problem of "two Oregons" divided by wealth, by economic prospects, and increasingly by world view. In 1960, for example, per capita income in affluent, suburban Washington County near Portland was 10 percent higher than the rest of the state. By the 1980s it was 25 percent higher. In the 1990s, problems of chronic unemployment and underemployment have been exacerbated by federal resource conservation policies. The Endangered Species Act and related policies have affected forest resources and fisheries. As both Matthew Slavin and Michael Hibbard point out, Oregon land use planning has been ineffective in responding to problems of rural economic decline.

A very specific issue resulting from the problems of the "other Oregon" has been the effort to define "secondary lands"—the less productive lands within rural areas. In 1974, Oregon policy makers assumed that two competing land uses—urbanization and resource production—needed to be balanced in a statewide system. There has been widespread concern that the resource land regulations are overly restrictive since some designated resource lands cannot support viable commercial farming, ranching, or forestry. James Pease describes the long and contentious history that caused the 1993 legislature to permit homes on lots that were created and owned before 1985, except on the most productive farm and forest land.

A larger issue is the question of the realistic future for Oregon's resource communities. In a study of Coos Bay, historian William Robbins (1988) has found an important discontinuity for the years from 1945 to 1975. In the midst of a normal pattern of booms and busts, these decades stand out as unusually prosperous. The disturbing message is that Oregonians whose ideas about the timber industry were formed between 1945 and 1975 have been influenced by the only substantial era of consistently high production. The intersection of new technologies (cutting and harvesting with gasoline engines) and a booming market in southern California allowed an entire generation of high-profit production at rates unsustainable in the long run. The painful adjustments of the last decade have been a return to the "normal" pattern of the nineteenth and earlier twentieth centuries.

Experts on the American West have responded to the general crisis of resource industries in different ways. Urban planner Frank Popper sees a continued disinvestment leading eventually to federal repurchase of depopulated lands for nature preserves (Popper and Popper 1987). This is the discontinuity of economic collapse. Ed Marston, editor of the Colorado-based *High Country News,* looks at the rural West and sees the spread of cosmopolitan ideas from lifestyle enclaves slowly modifying isolated rural communities (Marston 1989). His more optimistic sense of discontinuity sees a "reopening" or "resettling" of the western frontier as an archipelago of liveable communities dependent on recreation, retirement dollars, and electronically networked businesses.

It is important to explore the roles that land use planning can play in facilitating a successful transition for the "other Oregon." The driving force in this program is the protection of the forest and agricultural industries; it is a resource conservation program only to the extent to which conservation supports these industries. Dwellings are restricted or in some cases not allowed on resource lands for the purpose of managing for wildlife, for example; open space preservation and management are not considered to be resource uses. The connections between resource concerns have been little emphasized and as a result there is no clear understanding about the true costs of farm and forest practices. Deborah Howe touches on this concept in outlining a research agenda.

The "other Oregon" challenge has much to do with addressing conflicts between resource uses and rural development. While development on resource lands is restricted, there is, in fact, a considerable amount of rural development in what are known as exception zones: areas identified in acknowledged plans to be unsuitable for resource use due primarily to existing development patterns and parcel configuration. Many of these exception areas embrace well-defined and in many cases vital rural communities. The essays by James Pease and by Robert Einsweiler and Deborah Howe explain that the Oregon system essentially treats these communities as nonentities. They exist merely as exceptions to the urban and resource goals. Einsweiler and Howe argue strongly for reclassifying urban areas (incorporated cities) and rural communities as "human settlements." This would allow maintenance of the system's development/resource duality, which is justified as a response to market realities in which high density development can outbid resource land prices.

Critics have also pointed out that the LCDC system has had little to say about social equity issues. The goals use the qualifier " . . . consideration of the factors of environmental, social, and economic consequences." However, there has been far more attention to economic and environmental impacts than to the social effects of planning. Social systems are regarded, if at all, as reactive adapters to decisions driven by broadly economic imperatives. This silence about the social dimension is of particular concern in the light of demographic changes that saw a 24 percent increase in the state's African-American population, a 71 percent increase in its Hispanic population, and nearly a 100 percent increase in its Asian-American population during the 1980s.

The main arena in which the Oregon system has addressed social issues has been housing. Reflecting the strong interest during the 1970s in "fair share" housing policies that tried to distribute low-income housing throughout entire metropolitan areas, Goal 10 requires that jurisdictions provide "appropriate types and amounts of land . . . necessary and suitable for housing that meets the housing needs of households of all income levels." In an early assertion of its authority, LCDC forbade the small town of Durham in Washington County to shift its entire multifamily zone to single-family zoning. The City of Milwaukie ran into trouble by trying to set more stringent review standards for apartments than for detached houses. In 1982, the small suburban Portland municipality of Happy Valley became a test case when LCDC ordered it to plan for a substantially greater residential density than its residents desired.

In the last ten years, however, the Oregon system has viewed housing issues largely in terms of cost. Policy makers have asked whether urban growth boundaries raise housing costs by artificially restricting the supply of land, or lower them by promoting higher densities that support the efficient delivery of public services. As Nohad Toulan describes, however, empirical data show that a broad mix of affordable housing has been maintained in the Portland area. His chapter also places the Oregon system within the broad perspective of American efforts to deal with social equity issues through the planning process.

Other social issues are also worth noting. Citizen involvement in land use planning peaked in the 1970s, with the adoption of the statewide goals and comprehensive plans. For many Oregonians in the 1990s, planning is part of a bureaucratic routine rather than an active contributor to livability. Carl Abbott maintains that Oregon's planning style has tended to create bureaucratic procedures that operate fairly rather than arbitrarily. At the same time, bureaucratization reflects the declining role of citizen participation since the writing of statewide goals and local comprehensive plans in the 1970s. There is a need to reinvigorate public interest and involvement as new planning issues emerge, such as the current discussion about the most desirable forms of growth in the Portland area.

Another important planning issue on which the Oregon system offers little guidance is the needs of individuals with different levels and types of abilities, resources, and circumstances. The LCDC system essentially treats Oregonians as "economic persons" and places certain limits on their freedom within the market. It has been silent on the special problems that racial minorities may face in obtaining housing, that physically limited persons may have in reaching job sites, or that single parents may face in finding everyday services at convenient locations. In particular, the state system leaves the questions of scale and mix of land uses—one of the central planning concerns of the 1990s—to the discretion of local jurisdictions.

As John DeGrove points out, Oregon has been a model for other states. Oregonians have been key players in national communication networks on state planning, with staff of the land use advocacy group 1000 Friends of Oregon playing prominent roles. Several aspects of the state system have been especially exportable. These include its emphases on certainty and timeliness in procedures; its requirement of consistency between local plans and state standards; its use of urban growth boundaries; and its emphases on the protection of resource land and affordable housing.

In this light, Oregon has something to learn from other states. Its system began as "state-local conjoint planning" (Bollens 1992) rather than as the purely regulatory intervention characteristic of other early state efforts. A balanced growth focus, encompassing both environmental protection and accommodation of development, also emerged early. This emphasis now characterizes more recently adopted state programs as well (Bollens 1992). Yet this balanced system, which Frank Popper (1988, p. 297) has called "the most impressive case of political consolidation," is also one that continued to be dominated by the state (Gale 1992). The Oregon system might benefit from close attention to the more collaborative, consensual model that Bollens (1992) and Innes (1992) have described for New Jersey.

Indeed, DLCD has begun to recognize that Oregon in the 1990s can learn from second-generation planning efforts in other states. Serious consideration is being given to Florida's notion of concurrency, which requires that infrastructure be in place before development approvals are granted. Cross-acceptance, a process used by New Jersey to craft a consensus on the statewide plan, is being reviewed in Oregon as a means to achieve coordination among local government plans and policies within metropolitan regions. DLCD has also looked at other states for ideas on managing exurban development, state agency coordination, regional review of local plans, and state funding for infrastructure.

If the Oregonians who supported the state's planning program nearly twenty years ago had foreseen all of the challenges that lay ahead—the hard work, the opposition, the changing context—they might never have started. Fortunately, however, they were dedicated, idealistic, and perhaps naive. By the time the challenges arose, they had the requisite skills and confidence that could not have existed at the program's inception.

The program has evolved because Oregonians have been learning from their mistakes. This process is not always systematic, rigorous, or objective. It was not until 1990, for example, that DLCD undertook an evaluation of urban growth and forest and farm management. Political pressures and resource constraints have dominated and will continue to dominate the process of change, but the basic tenets of equity in decision making, resource protection, and community vitality continue to serve as the program's guiding principles. The extent to which Oregon can foster a culture of learning will determine the relevance of the statewide planning program as a framework for meeting the needs of the twenty-first century. It is a commendable system and well worth efforts to adjust, refine, and improve.

Notes

1. The following legislative history is based in part on an interview with Hector Macpherson, Ted Hallock, Stafford Hansell, and Henry Richmond, conducted at the Oregon Historical Society by Carl Abbott and Deborah Howe, December 14, 1992 (Abbott and Howe 1993).
2. The emphasis on farmers' property rights has created serious problems in selling mandatory planning in other states.
3. The LCDC has been chaired in chronological order by the following: L. B. Day (Salem); John Mosser (Portland); Richard Gervais (Bend); Lorin Jacobs (Medford); Stafford Hansell (Hermiston); Stanton Long (Eugene); and William Blosser (Dayton). The directors of DLCD have been Arnold Cogan, Harold Brauner, Wes Kvarsten, James Ross, Susan Brody, and Richard Benner. Senate Bill 100 also created a permanent Joint Legislative Committee on Land Use to advise LCDC, review its actions, and recommend needed legislation.
4. Outside the Willamette Valley, editorials favorable to the state system appeared in the Bend *Bulletin* (October 16, 1978); Klamath Falls *Herald and News* (October 26, 1978); Medford *Mail-Tribune* (October 10, 1978); Pendleton *East Oregonian* (October 13, 1978); Redmond *Spokesman* (October 11, 1978).
5. Editorials favorable to the state system in 1982 included the Bend *Bulletin* (October 17); Hood River *News* (October 13); *Newport News-Lincoln City Times* (October 13); Medford *Mail-Tribune* (October 27); Pendleton *East Oregonian* (October 20); Salem *Statesman-Journal* (October 24); Eugene *Register-Guard* (October 10); and Portland *Oregonian* (October 19). Opposition was found in the Baker *Democrat-Herald* (October 12) and the Klamath Falls *Herald and News* (October 18).

References

Abbott, Carl, and Deborah Howe. 1993. "The Politics of Land-Use Law in Oregon: Senate Bill 100 Twenty Years After." *Oregon Historical Quarterly*, 94, 1: 1-35.

Bollens, Scott. 1992. "State Growth Management: Intergovernmental Frameworks and Policy Objectives." *Journal of the American Planning Association*, 58, 4: 454-466.

Gale, Dennis. 1992. "Eight State-Sponsored Growth Management Programs: A Comparative Analysis." *Journal of the American Planning Association*, 58, 4: 425-39.

Innes, Judith Eleanor. 1992. "Group Processes and the Social Construction of Growth Management: Florida, Vermont, and New Jersey." *Journal of the American Planning Association*, 58, 4: 440-53.

Marston, Ed. 1989. *Reopening the Western Frontier*. Washington: Island Press.

Popper, Deborah Epstein, and Frank J. Popper. 1987. "The Great Plains: From Dust to Dust." *Planning*, 53, 12: 12-18.

Popper, Frank J. 1988. "Understanding American Land Use Regulation Since 1970: A Revisionist Interpretation." *Journal of the American Planning Association*, 54, 3: 291-301.

Robbins, William G. 1988. *Hard Times in Paradise: Coos Bay, Oregon, 1850-1986*. Seattle: University of Washington Press.

Vogel, David. 1985. *National Styles of Regulation*. Ithaca: Cornell University Press.

PART I
Building the Oregon System

CHAPTER 1
Land Use Politics in Oregon

Gerrit Knaap

Oregon is widely viewed as enigmatic. People east of the Rocky Mountains know that Oregon lies somewhere north of San Francisco, with Seattle as its capital city. They also know that Oregon has border guards, and that it rains nearly every day. What's more, they know that Oregon is the place where land use issues have been resolved long ago by locking up land through firmly established and permanent land use controls (*Wall Street Journal* 1982).

While national perceptions concerning the relative abundance of rain in Oregon may be essentially correct, perceptions concerning land use issues in Oregon are not. Land use controls in Oregon are not cast in stone, land has not been locked up, and land use issues are far from resolved. Oregon does have a unique land use program, but the program is characterized less by stability and harmony than by conflict and change.

Oregon's land use program includes goals and local comprehensive land use plans and regulations to implement those plans. The program is perhaps the most ambitious and highly acclaimed land use program in the nation. But in Oregon, as in all other states, land use plans and regulations continually change, and many conflicts over land use remain unresolved. Oregon's land use program thus represents not the resolution of land use conflicts but a political process through which land use conflicts can be resolved.

What is different about land use in Oregon is the intergovernmental process through which land use decisions are made. Like most other states, Oregon enabled local governments to plan and zone land use before 1973. In that year, however, land use planning and zoning became more than local opportunities. Following the passage of Senate Bill 100, local governments in Oregon *must* plan in a manner consistent with *state* land use goals and guidelines. If local governments fail to plan and regulate land accordingly, as determined through an "acknowledgment" and review process,[1] the state can preempt local land use authority and withhold state grants and aids. As a result, land use planning and regulation in Oregon involves state *and* local land use politics.

In this chapter I discuss land use politics in Oregon. I structure the discussion jointly by steps in the planning process (e.g., program adoption, policy formation, plan preparation, plan acknowledgment, and plan implementation), and by the participants in the political process (e.g., Oregon citizens, the Oregon legislature, state-level interest groups, local interest groups, state agencies, and local governments). I proceed as though the steps in the planning process occur sequentially and are shaped by political conflict between successive (though not mutually exclusive) sets of participants. Land use politics and the process of land use planning and regulation are not, of course, so neatly segmented. New land use bills are continually enacted in response to changing social and economic circumstances, and plans are continually changed throughout the process of plan implementation. Further, Oregon's citizens, legislature, interest groups, state agencies, and local governments are to some extent involved in every step of the planning process. But, as I argue in the following pages, the politics of land use in Oregon are dominated by different forms of political conflict at different stages in the planning process; what's more, these differences increase the difficulty of maintaining consistency among land use programs, land use policy, land use plans, land use regulations, and land use.

The Politics of Reform

Until 1973, the politics of land use in Oregon resembled those in other states (Bureau of Government Research and Service 1984). Participants in the process included developers, local residents, and city and county governments. In some cities and counties, land use was planned. But because plans were not legally binding, planning attracted little political interest. Instead, political conflicts developed over zoning, subdivision controls, public works, and other public activities with immediate effects on land use. The outcome of conflicts between residents and developers varied from time to time and from place to place.[2] But by the late 1960s no-growth sentiments tended to prevail in urban areas and in the mature suburbs, and pro-growth sentiments in rural areas and in the unincorporated urban fringe. Local land use governance thus fostered urban decentralization and rapid development of farm and forest lands.

National social and economic trends during the 1950s and 1960s, and their manifestations in Oregon, created discontent with the existing structure of land use governance and political momentum for reform (Leonard 1983, Knaap 1987b). Rapid western migration spilled over the California border and into Oregon's interior valleys. Economic transformation reduced the demand for farm and forest workers and increased the demand for skilled professionals in light manufacturing, services, and retail trade. Rising incomes and falling transportation costs enabled urban workers to commute from mini-farms and ranchettes to jobs in Portland, Salem, Eugene, Medford, and Bend. Combined, these trends made farm land in the Willamette Valley more valuable to urban commuters than to farmers, forests more valuable for recreation than for timber, and urban residents more interested in urban growth management than in urban growth. Tom McCall's now-famous plea for land use reform expressed the mood of a growing number of Oregonians:

> There is a shameless threat to our environment . . . and to the whole quality of life—[that threat] is the unfettered despoiling of the land. Sagebrush subdivision, coastal "condomania" and the ravenous rampage of suburbia in the Willamette Valley all threaten to mock Oregon's status as the environmental model for the Nation. We are in dire need of a state land use policy, new subdivision laws, and new standards for planning and zoning by cities and counties. The interest of Oregon for today and in the future must be protected from the grasping wastrels of the land (quoted in DeGrove 1984, p. 237).

Tom McCall's speeches did much to fuel the reform movement. With the support of a popular governor, the legislature in 1973 passed a pioneering land use bill which transferred much of the power to control land use from the local to the state level. Around the same time the Oregon legislature also passed laws placing deposits on beverage containers, prohibiting billboards along scenic highways, and protecting beaches from private development. Oregon's land use program subsequently became regarded as another expression of Oregon's renegade brand of environmentalism. But such a view misses the subtleties of land use politics in Oregon.

Although popular support for reforming land use governance in the 1970s was widespread and growing, there was still conflict. Initiative petitions to end state participation in land use control placed the issue on a ballot measure four times (1970, 1976, 1978, and 1982). Each time the proposal was defeated. But analyses of the results of these ballot measures found popular support for state land use reform sharply divided by region, by occupation, and by residential location (Medler and Mushketel 1979, Knaap 1987a, 1987b).[3]

Popular support for statewide reform was perhaps most sharply divided between regions: support for reform was greater in the Willamette Valley than in other regions of the state—for rather obvious reasons. The Willamette Valley was the most rapidly growing region in the state; and, as the population grew, more and more farm land developed into subdivisions that resembled southern California—a resemblance many Oregonians sought to avoid (DeGrove 1984). Urban development in the Willamette Valley seemed beyond the capabilities, or interests, of local governments to control. From the perspective of Willamette Valley residents, especially those in urban areas, state land use reform offered a solution to the failure of local governments to control urbanization in the Willamette Valley.

To residents of other parts of the state, however, statewide reform was much less attractive. Although population growth also sparked land use conflicts in eastern Oregon, in the Cascade Mountains, and along the Pacific Coast, conflicts in these regions were less pervasive and not beyond the abilities of local governments to control. Weak government control over development outside the Willamette Valley reflected local desires for less intervention rather than a lack of capacity to intervene. Further, the state capital stood in Salem, a location far in distance and culture from the timberlands of Roseburg, the rangelands of Prineville, and the fishing docks of Gold Beach. To residents of these areas, then, land use reform threatened to transfer control over their land to bureaucrats in Salem.

Popular support for reform was also divided by occupation. Whereas Oregonians in the trade, service, and communications industries supported reform, those in the construction, farming, and forest-products industries opposed it. This division is also easily understood. Statewide land use reform, as it developed in Oregon, promised environmental protection and resource conservation. For those whose livelihoods were not tied to the resource base, such a promise offered "environmental" income—i.e., income in the form of open space, pristine forests, and scenic ocean beaches (Whitelaw and Niemi 1989). For those whose livelihoods were dependent on resource extraction and development, statewide reform threatened jobs in home construction, farming, and forestry.

Finally, popular support for reform was divided by residential location. Urban residents supported reform; rural residents did not. To some extent, this division reflected that between regions and occupations, since urban residents are more likely to live in the Willamette Valley and work in nonresource-based occupations. But it also reflected a difference in interests independent of differences between regions and occupations: a difference over development rights. Whereas urban residents typically own developed land, rural residents typically own undeveloped land. And since statewide reform promised to control urban development, such reform was significantly more attractive to owners of already developed land.

Social and economic trends in Oregon during the 1950s and 1960s thus tipped the balance of popular politics in favor of land use reform. Rapid population migration enabled urban residents in the Willamette Valley to dominate state-level popular politics. Development beyond the control of local governments in the Willamette Valley created a demand for centralized land use governance. Economic transformation caused urban residents to favor resource conservation over development. And growing home ownership in Oregon's cities strengthened support for controlling rather than stimulating urban growth. These social forces created political momentum for reform which featured centralized control, resource conservation, and urban growth management.

The Politics of Program Adoption

Popular pressures for land reform in the late 1960s and early 1970s were not restricted to Oregon. Oregon was only part of a "quiet revolution" that swept the nation during this period (Bosselman and Callies 1971, Popper 1981), inspiring a variety of land use reforms. Some states required local governments to plan and regulate land use; others required state governments to regulate select areas and developments of a certain scale; still others required state governments to plan and regulate all land use (Rosenbaum 1976). Popular pressure thus caused many states to reform the structure of land use governance, but the new form of governance structure was crafted by state legislatures and thus by legislative politics.

Oregon's legislature is relatively weak (Hedrick and Zeigler 1987). Legislators in Oregon have short terms, meet only once every two years, are poorly paid, and have limited staff. Political parties in Oregon are also weak. Direct primaries force Oregon legislators to seek financial support from local citizens and interest groups and to establish political organizations outside the state party machinery. As a result, state political candidates come to the legislature on local, not on partisan platforms.

Due in part to the weakness of political parties and in part to Oregon's until recently relatively undiversified economy, political interest groups are active and influential in the Oregon legislature (Hedrick and Zeigler 1987). They represent utilities, health and medical organizations, education, financial institutions, the building and construction industry, business in general, and local governments. Contrary to popular perceptions, political influence in Oregon is not dominated by organizations representing agriculture and wood products but instead by organizations representing utilities and "new wave" business interests (e.g., Tektronics, Intel, Hewlett-Packard, and Nike).

Following the American Law Institute's Model Land Development Code, and the trend in other states (Rosenbaum 1976), the original version of Senate Bill 100 would have greatly expanded state land use authority via state-level permit authority, state-controlled areas of critical concern, and regional land use councils—enough state authority to draw the opposition of nearly all interest groups. Facing certain defeat in the Senate Environmental and Land Use Committee, Senate Bill 100 was referred to an ad hoc committee, whose charge was to produce a bill that could be passed by the committee and the entire legislature.

The bill that came out of committee, and was subsequently passed by the legislature, established a novel statewide land use program. The bill required local governments to formulate comprehensive plans; it created a state land use agency, the Land Conservation and Development Commission (LCDC) and its administrative arm the Department of Land Conservation and Development (DLCD); and it required LCDC to assure that local plans meet state land use goals through an acknowledgment process. But the bill contained no specific rules for governing the review process; it excluded state permitting authority and regional land use councils; and it left the substance of Oregon's land use policy undetermined. Thus, in spite of a public mandate for change, the Oregon legislature severely limited the extent and specificity of land use reform.

The structure of Oregon's land use program vividly reflects the influence of legislative politics. Due to the weakness of Oregon's political parties, Senate Bill 100 was not ushered through the legislature on the platform of the dominant political party. Instead, it survived the legislative process only after being pillaged by interest-group politics. The bill survived because it received the support (or failed to draw the opposition) of Oregon's most powerful interest groups: the utilities and the "new wave" industries. But the dependence of legislators on support from local constituents weakened the extent of state intervention and preserved a considerable degree of local control. Even then, legislative votes on the bill were sharply divided by location: legislators from the Willamette Valley voted forty-nine to nine in favor; legislators from all other regions voted twenty-one to nine against (Little 1974).

The Politics of Policy

Though perhaps necessary for passage through the legislature, the ambiguity of Senate Bill 100 left many policy issues unresolved. What should be the goals of the state land use program? How would such goals be interpreted? Which goals should receive priority? These issues were resolved through the politics of state land use policy.

Although the legislature provided some general guidelines, Oregon's statewide land use goals were established by the newly formed state land use agency, LCDC, as standards for reviewing comprehensive plans and for guiding land use decisions before comprehensive plans were acknowledged.

The adoption of statewide goals and guidelines, however, did not end the policy formation process, and LCDC did not establish state land use policies alone. Several of the goals conflicted, and the goals themselves offered little substance with which to evaluate local land use plans. Specific state land use policies had to be established through the acknowledgment process (discussed below), through legislative oversight, and through judicial review. Although these processes were led by elected and appointed officials, state-level interest groups played an important role. Interest groups active in the politics of state land use policy differed from those active in the politics of program adoption. In general, participants in the politics of land use policy making included those interest groups most directly affected, such as the development industry (including the

Oregon Aggregate and Concrete Producers, the Oregon and Portland Home Builders Associations, and Associated Oregon Industries) and environmental organizations (especially 1000 Friends of Oregon) (Liberty 1989).

Perhaps the most influential interest group in state policy decision making is 1000 Friends of Oregon, which was formed as an independent watchdog organization to give "the people of Oregon a powerful tool to help America's leading state land use program succeed" (1000 Friends of Oregon, 1982a). Funded by donations, gifts, foundation grants, and dues from its 5,000 members (Liberty 1989), 1000 Friends has been an active lobby in the state legislature, a regular participant in the acknowledgment process, and a frequent instigator of precedent-setting judicial reviews.

Often in response to pressure from 1000 Friends and other interest groups, the legislature played a major role in establishing state land use policy, both by amending Senate Bill 100 and by appropriating funds for the planning process. Some examples are as follows. In 1977, the legislature repealed LCDC's authority to enact and enforce its own plan for recalcitrant local governments and authorized LCDC to adopt enforcement orders;[4] the 1979 legislature created the Land Use Board of Appeals;[5] the 1981 and 1983 legislatures established and revised the post-acknowledgment review process;[6] the 1983 legislature required LCDC to place more emphasis on economic development and public facilities; the 1985 legislature required LCDC to study means for designating secondary farm land;[7] the 1987 legislature created the Ocean Resources Planning Program and removed from counties control over forest practices; and the 1989 legislature appropriated funds to study the effectiveness of urban growth management and farm land preservation. In sum, the legislature altered some aspect of the program in nearly every legislative session.

To monitor the program between legislative sessions, the legislature created a Joint Legislative Committee on Land Use. Proposals for land use legislation, including funding proposals, often originated in this committee. The legislature appropriated an average of $3.6 million each year to administer the program. From 1973 to 1989, 56 percent of state appropriations went to local governments for preparing plans and for plan implementation—largely without strings attached (Oregon DLCD 1991a). Less than half the appropriations went to DLCD for staffing reviews, monitoring, and research. This pattern of appropriation curtailed the influence of LCDC and preserved the influence of local governments over local land use.

As the legislature enacted new land use laws, Oregon's appellate courts were actively interpreting them. These interpretations helped establish the substance as well as the process of land use planning. The Oregon courts ruled, for example, that agricultural land goals dominate housing goals outside urban growth boundaries (UGBs) (*Peterson v. City of Klamath Falls*), that zoning urban land for low-density use can violate state housing goals (*Seaman et al. v. City of Durham*), and that the capacity of public services can be considered in changes to land use plans or regulations (*Dickas v. City of Beaverton*). With these and other precedent-setting interpretations, the Oregon courts thus established key elements of state land use policy.

Through the process of administration, legislation, and adjudication, Oregon's land use goals and policies became codified into specific and binding administrative rules, land use statutes, and case law—often at the instigation of state-level interest groups. As a result, those goals that attracted the attention of state-level interest groups (e.g., urban growth management, housing, farm and forest land protection) dominated the planning agenda; those goals without an active state constituency (e.g., energy conservation, recreation, and natural hazards) received little attention (Liberty 1989). In essence, during policy formation a compromise between development and environmental interests, the major interest-group participants, emerged. Comprehensive plans had to permit—in fact encourage—urban development inside urban areas while protecting farm and forest land from development outside urban areas.

The Politics of Planning

Although the statutory structure and policy thrust of Oregon's land use program were established at the state level, responsibility for the most fundamental aspect of the program remained at the local level: local governments had to prepare comprehensive land use plans. And while the process of policy formation took place in Salem, the state capital, land use plans were prepared in city halls and county seats around the state, in local political environments.

Although planning continued to take place at the local level, the politics of local planning changed in 1973. Before 1973 local governments could choose to plan and could plan to pursue any locally chosen land use goal. After 1973 local governments *had to* plan, and do so in accordance with specific *state* land use goals and guidelines. Further, the Oregon Supreme Court ruled in 1973 that acknowledged plans were controlling instruments, with which all local land use decisions had to conform (*Fasano v. Washington Co.*). These structural changes in land use governance enabled state agencies and interest groups to influence the content of local land use plans and created local interest in land use planning where before there was interest only in land use regulation.

But in spite of state-prescribed procedures and goals, local plans continued to reflect local politics, which in urban areas continued to be dominated by developers and homeowners. Under pressure from homeowners, for example, municipal governments still sought to limit growth through exclusionary zoning. And under pressure from developers and business interests, municipal governments still sought to facilitate development—typically at the urban fringe. As a result, plans prepared by municipal governments featured low-density zoning and extensive UGBs.[8]

Politics in rural areas meanwhile continued to be dominated by farmers and farm organizations. As a result, county governments sought to maintain farm productivity by containing urban growth. County governments also sought to protect farm land values by imposing few restrictions on land partitions and building permits. As a result, plans prepared by county governments favored urban growth containment and little protection of farm and forest land.

The difference in political environments between municipal and county governments created intergovernmental conflicts—especially over UGBs. Before a UGB could be submitted to LCDC for acknowledgment, all affected local governments had to agree on the placement of the boundary. This forced municipal and county governments to settle through the planning process conflicts that would otherwise have occurred through the gradual process of municipal annexation and unincorporated urban growth. Although the process took years and extensive prodding from LCDC, UGBs were eventually established. But in many cases agreement could only be reached by making special exceptions, contingencies, and stipulations.[9]

The Politics of Acknowledgment

After their plans had been prepared, local governments had to submit their plans to LCDC for acknowledgment review. During the review process, LCDC determined whether the plan submitted by a local government complied with state land use policies. Plans that did not comply had to be revised until they did. Plans that did comply were acknowledged. The key players in the acknowledgment process included local governments, who submitted plans; LCDC, who reviewed the plans; and state-level interest groups, who also reviewed the plans and submitted written comments. Once again the process created intergovernmental conflicts—this time between state and local governments. Although the process was originally scheduled to finish in 1975, the last plan was acknowledged in 1986, more than ten years after the process had begun.

Intergovernmental conflict over acknowledgment was endemic to the process. After all, the statewide planning program had been created out of dissatisfaction with local planning programs. And state land use goals and policies were formed by organizations with statewide interests, while land use plans were produced by organizations with local interests. The acknowledgment process thus served as the principal forum for addressing conflicts between state and local interests over land use.

One of the first issues of contention in the acknowledgment process was how much land to allow inside urban growth boundaries. According to the state land use goals and guidelines, all urban areas in the state had to be encircled by UGBs, within which all urban development must take place. Local governments submitted plans with large UGBs; 1000 Friends and other state environmental groups wanted small UGBs.[10] During the acknowledgment process, DLCD often sided with 1000 Friends, and required many local governments to reduce the size of their proposed UGBs (Knaap 1991).

A second contentious issue concerned zoning for residential use. Local governments submitted plans with extensive low-density zoning and placed high-density housing in conditional use zones (thereby allowing the local government to reject proposals for high-density housing unless the proposal met stringent and often ambiguous conditions). Developers and contractors, with the support of 1000 Friends, wanted high-density zoning and clearly established criteria for high-density development. Again, DLCD often sided with 1000 Friends and the developers and forced nearly every local government to increase the density of residential zoning and to clarify the conditions for high-density land use (1000 Friends 1982b, Knaap 1991).

A third issue concerned the protection of farm and forest land. Local governments, especially county governments, submitted plans with weak and ambiguous standards for development in farm and forest zones. 1000 Friends wanted strict standards—standards based on farm and forest performance. Once again, LCDC often sided with 1000 Friends, and forced county governments to adopt strict and explicit standards for development outside UGBs (Leonard 1983).

As these issues illustrate, the acknowledgment process had substantive impacts on the content of local comprehensive plans. With some exceptions, the acknowledgment process resulted in tighter urban growth boundaries, tighter development restrictions in farm and forest zones, and looser development restrictions in residential zones. Through the acknowledgement process the compromise between development and environmental interests—encouraging development inside UGBs and discouraging development outside UGBs —became incorporated into local comprehensive plans.

The Politics of Implementation

After a local government had its plan acknowledged by LCDC, it then had to implement its plan, a process which continues today. Although local governments were charged with both preparing and implementing their plans, the politics of plan implementation differ from the politics of plan formulation because the nature of land use decisions, the effects of decisions, and the participants in the decision-making process all differ.

The nature of the land use decisions made during plan implementation differ considerably from those made during plan formulation. A decision to zone land for industrial use, for example, differs substantially from a decision to grant a building permit for an industrial plant—even though the plant meets the criteria established for the industrial zone. The decision differs in two critical respects. First, the decision to permit an industrial plant will result in a short-term or immediate land use change, whereas the decision to zone land for industrial use may never result in land use change. Second, the decision to permit a plant is made with much greater information than the decision to zone land for industrial use. When an application for a building permit is submitted, information is often available on the size of the plant, employment, and possible environmental effects. Information available during plan implementation might make the plant more or less attractive, but this new information could easily cause local governments to reconsider decisions made during plan formulation.

The effects of land use decisions made during plan implementation differ in distribution from those made during plan preparation. Decisions made when formulating plans affect categories of interests and persons; decisions made when implementing plans affect specific interests and persons. A decision to zone land for industrial use, for example, benefits industrial interests and perhaps industrial workers; a decision to grant a building permit benefits a particular corporation and perhaps its prospective employees. The benefits of an implementation decision, therefore, are concentrated and potentially quite large. By similar logic, the costs of an implementation decision are also concentrated and equally large. As a result, firms or individuals might be willing to commit considerably greater resources to influence decisions made during plan implementation than they would during plan preparation.

Finally, participants in the process of plan preparation differ from those who participate in the process of plan implementation. Participants in the process of preparing plans often include organized groups of developers and homeowners with a collective and long-term interest in local land use. Participants in the implementation process often include individual developers and landowners, including those who care little about land use planning but much about zoning and new developments. All these differences contribute to an "implementation deficit" between land use plans and land use decisions (Downing 1984).

But the implementation deficit between land use plans and plan implementation in Oregon stems in part from an additional source: the intergovernmental structure of its land use program. Through the process of acknowledgment, statewide interest groups and agencies are able to influence the content of local land use plans, but no similar process enables them to influence plan implementation. Implementation takes place on a piecemeal, long-term, day-to-day basis, as local governments construct roads, extend sewers, approve subdivisions, enforce zoning ordinances, and grant building permits. State interest groups and agencies cannot, therefore, possibly participate in all the land use decisions involved in the process of plan implementation.

The difference in politics between plan preparation and plan implementation threatens to undermine the policy framework established by the compromise between state-level interest groups. This threat, perhaps the most pervasive land use issue in Oregon today, recently stimulated extensive research on the effectiveness of local implementation. The evidence confirms there is cause for concern. Since 1981, for example, counties have approved between 85 and 96 percent of all applications for land divisions and new dwellings in exclusive farm use zones (Liberty 1989). And the majority of building permits and land subdivisions, contrary to state land use goals, are not being approved to enhance commercial farming (Oregon DLCD 1991b). Further, development has occurred at densities less than planned inside UGBs and at densities greater than planned outside UGBs (DLCD 1991c). In short, there is evidence that land use goals formed at the state level are being systematically undermined by land use decisions made at the local level.

The Politics of Land Use in Oregon

In sum, land use issues in Oregon are far from resolved and remain subject to land use politics. The politics of land use in Oregon occur in different venues at different stages of the planning process. A simple model of the process is illustrated in figure 1. The model in the figure lists the participants in state land use politics, including agencies of state and local governments and land use constituencies ranging from the entire Oregon citizenry to individual landowners. Each triangle in the model depicts a particular stage in the planning process; from left to right the stages include program adoption, policy formulation, plan acknowledgment, comprehensive planning, and plan implementation.[11]

Although a rather simple heuristic, the model illustrates some important features of state land use politics in Oregon. First, the model illustrates how different organizations and agencies participate in different stages of the planning process. As depicted here, Oregon's land use program was conceived by popular pressure for reform and shaped by interest group politics in the legislature. Once adopted, the program's policy framework was established by the legislature, by the institutions of state government (especially LCDC, DLCD and the state courts), and by interest groups with statewide interests in land use planning and regulation. State land use policies were incorporated into land use plans through the process of acknowledgment, a process dominated by state agencies, state interest groups, and local governments. Land use plans were prepared, however, by local governments, under pressure from local interest groups and according to parameters set by state agencies. Finally, plans are being implemented by local governments through everyday land use decision making and through regulations they impose on land owners.

Figure 1 also illustrates some built-in obstacles to maintaining a consistent set of policy objectives throughout the stages of planning, especially between plan preparation and plan implementation. Oregon's land use program was designed to further state land use goals while maintaining local control over plan implementation. But by maintaining local control, the program enables local governments to undermine—at least partially—state land use goals and objectives. The extent to which this occurs depends in part on the similarity of political forces that bear at the state level with those at the local level. Congruence between state and local politics is more likely, for example, in those regions of the state with strong popular support for statewide land use planning—e.g., in the urban areas of the Willamette Valley; less congruence is likely in the rural areas of eastern Oregon.

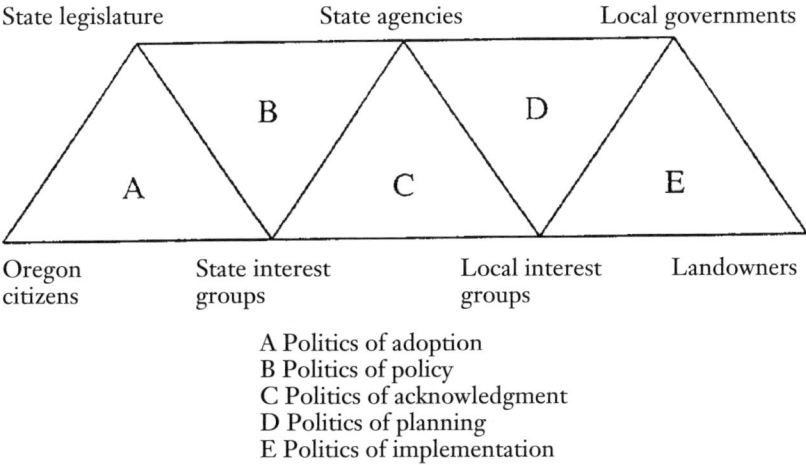

A Politics of adoption
B Politics of policy
C Politics of acknowledgment
D Politics of planning
E Politics of implementation

Figure 1. The participants in state land use politics.

As the program is currently structured, local governments hold the key to linking state-acknowledged plans and local plan implementation. The extent to which land use will eventually reflect state land use goals will depend primarily on the decisions made by local governments. Because comprehensive plans are legally binding policy instruments, however, local decision making is constrained. Local decisions inconsistent with acknowledged plans can be appealed to the Land Use Board of Appeals (LUBA). The criteria for standing to appeal are fairly liberal. But if judicial review became common for assuring consistency between land use decisions and land use plans, the program would quickly collapse under its own weight.

At present, efforts to assure that local implementation does not undermine state land use policies are proceeding in three directions. One is to impose greater state control over land use decision making. Another is to encourage greater intergovernmental cooperation between state and local governments. A third is to seek greater political integration between state and local politics.

Perhaps not surprisingly, the movement to impose greater state control over implementation is being led by state-level interest groups—especially 1000 Friends and the Portland Metropolitan Homebuilders. Based on research that found the density of residential development in the Portland metropolitan area far below planned designations, 1000 Friends and the Metropolitan Homebuilders advocate rules requiring local governments to impose minimum density standards for development approval. They also advocate rules requiring local governments to monitor urban growth patterns and to demonstrate compliance with regional housing objectives (1000 Friends and the Metropolitan Homebuilders 1991). 1000 Friends has also introduced proposals to substitute state for county administration of building permits, land divisions, and other land uses in farm and forest zones (Liberty 1988). At present, however, legislative support for these proposals remains uncertain.

The movement toward greater state and local cooperation is led by DLCD. In the post-acknowledgment period, DLCD must review the plans of local governments on a periodic basis to assure that local plans continue to comply with state land use policy. And, through the process of plan amendment, DLCD helps local governments continually amend their plans to meet changing local conditions and new state policy initiatives. In the first year of post-acknowledgment, DLCD received about three thousand proposed amendments; it participated in about 40 percent of those and appealed seven to LUBA (Oregon DLCD 1991a). Most often DLCD advised local governments how plan amendments could be made consistent with state planning goals. Thus, by working closely and continuously with local governments, DLCD is striving to maintain consistency between state land use policy and local land use plans.

To further consistency between local plans and local implementation, LCDC is responsible for coordinating reviews of programs of state agencies that affect land use. Because road construction, public parks, sanitary services, and other critical elements of plan implementation are financed in large part from intergovernmental grants, the Department of Transportation, the Department of Parks and Recreation, the Department of Environmental Quality, and 23 other state agencies exercise considerable influence over plan implementation. Thus, by reviewing the programs of state agencies, LCDC can ensure that those aspects of land use decision making influenced by state agencies are consistent with state-acknowledged plans.

Finally, several interest groups have begun new initiatives in both state and local political arenas. As demonstrated by 1000 Friends, state interest groups can be effective in multiple political arenas by educating the public, by lobbying the legislature, by conducting policy research, and by monitoring the activities of local governments. Recently, 1000 Friends has begun to form local networks for influencing planning and plan implementation at the local level. On the opposing side, farm organizations are beginning to move beyond politics at the local level and to increase their political activity at the state level. Oregonians in Action, who oppose many aspects of land use planning as interference with property rights, for example, recently emerged with a paid staff, legal council, a newsletter, and an active lobbying campaign (Pease 1990).

In sum, links are developing to integrate the various political arenas and stages of state land use planning. Some links will develop in the form of new rules and regulations, others will stem from greater intergovernmental coordination, and still others will reflect greater interest-group participation in multiple political arenas. The extent to which these links will overcome systemic obstacles to integrating planning and plan implementation, however, remains to be seen.

Closing Comments

In Oregon's future, the cycle of legislative action, policy formulation, planning, plan review, and plan implementation is likely to shorten considerably. No longer should it take nearly fifteen years for a popular mandate to alter local land use decision making. With acknowledged comprehensive plans for the entire state currently in place, and with new procedures for intergovernmental coordination, popular mandates for change will affect local land use plans much more quickly. The extent to which local land use decision making will further state land use policies depends on further progress at political integration throughout the stages of planning—especially between plan preparation and plan implementation. This complex, intergovernmental process of land use decision making represents, though, another uniquely Oregonian resource—one that can still be cherished while viewing scenic Willamette Valley farm land in the rain.

Notes

1. Formally, "acknowledgment of compliance with statewide planning goals."
2. For more on local land use politics see Davis (1963), Molotch (1976), Fischel (1985), and Johnson (1989).
3. These referenda were actually held after Oregon's program was adopted. But since the results of the referenda were similarly divided, the results are likely to reflect differences in popular support before the program was adopted.
4. Enforcement orders are a temporary suspension of local land use powers which LCDC can impose to compel a local government to make progress toward meeting one or more of the statewide planning goals.
5. The Land Use Board of Appeals is a special appellate court which hears only land use cases.
6. Under the revised post-acknowledgment review procedures, LCDC reviews the plans of local governments every four to seven years to assure that the plans remain in compliance with statewide planning rules.
7. "Secondary lands" are lands located in rural areas but poorly suited for farming or forestry.
8. Urban growth boundaries (UGB) are lines drawn around urban areas within which all urban development must take place. All land outside UGBs is designated for rural use unless specifically exempted.
9. In order to reach agreement on UGBs, some land outside the UGB was excepted from exclusive farm or forest use, some land inside the UGB was identified as "agriculturally soft" and thus protected from immediate development pressures, and some local plans were only partially acknowledged.
10. 1000 Friends held to the language in the goals which stated that UGBs must contain only sufficient supplies of land to meet the 20-year requirements for urban land use.
11. Figure 1 is an adaptation of the "iron triangle" between a legislative committee, an administrative agency, and a special interest group. See Moorehouse (1983).
12. For more on political integration and intergovernmental cooperation in state land growth management, see Innes (1991a, 1991b).

References

Bosselman, F., and D. Callies. 1971. *The Quiet Revolution in Land Use Control.* Washington: U.S. Council on Environmental Quality.

Bureau of Governmental Research and Service. 1984. *Guide to Local Planning and Development.* Eugene: University of Oregon.

Davis, Otto. 1963. "Economic Elements of Municipal Zoning Decisions." *Land Economics* 39, 4: 375-86.

DeGrove, John M. 1984. *Land Growth and Politics.* Chicago: Planners Press, American Planning Association.

Downing, Paul B. 1984. *Environmental Economics and Policy.* Boston: Little Brown.

Fischel, William. 1985. *The Economics of Zoning Laws.* Baltimore: Johns Hopkins University Press.

Hedrick, William, and L. Harmon Zeigler. 1987. "Oregon Politics of Power." In *Interest Group Politics in the American West,* edited by Ronald Hrebenar and Clive Thomas. Salt Lake City: University of Utah Press.

Innes, Judith. 1991a. "Implementing State Growth Management in the U.S.: Strategies for Coordination." Prepared for publication in *Growth Management and Sustainable Development,* edited by Jay Stein. Beverly Hills: Sage.

———. 1991b. "Group Processes and the Social Construction of Growth Management: The Cases of Florida, Vermont, and New Jersey." Paper presented at the Joint International Conference of the American Collegiate Schools of Planning and the Allied European Schools of Planning, Oxford, England.

Johnson, William. 1989. *The Politics of Planning.* New York: Paragon House.

Knaap, Gerrit J. 1991. "State Land Use Planning and Inclusionary Zoning: Evidence From Oregon." *Journal of Planning Education and Research* 10, 1: 39-46.

———. 1987a. "Self-Interest and Voter Support for Oregon's Land Use Controls." *Journal of the American Planning Association* 53, 1: 92-97.

———. 1987b. "Social Organization, Profit Cycles, and Statewide Land Use Controls in Oregon: Welcome to Oregon, Enjoy Your Visit." *Journal of Applied Behavioral Science* 23, 3: 371-86.

Leonard, H. Jeffrey. 1983. *Managing Oregon's Growth.* Washington: The Conservation Foundation.

Liberty, R. L. 1988. "The Oregon Planning Experience: Repeating the Success and Avoiding the Mistakes." Paper presented at the Conference on the Chesapeake Bay Critical Area Protection Program.

———. 1989. "An Overview of the Oregon Planning Program." Paper presented at the Lincoln Institute of Land Policy, Cambridge, MA.

Little, Charles E. 1974. *The New Oregon Trail.* Washington: The Conservation Foundation.

Medler, J., and A. Mushketel. 1979. "Urban-Rural Class Conflict in Oregon Land-Use Planning." *Western Political Quarterly* 32, 3: 338-49.

Molotch, Harvey. 1976. "The City as a Growth Machine: Toward a Political Economy of Place." *American Journal of Sociology* 82, 2: 309-22.

Moorehouse, Sarah McCally. 1983. *State Politics and Policy.* New York: CBS College Publishing.

1000 Friends of Oregon. 1982a. *Report for the Seventh Year: 1975-1982.* Portland: 1000 Friends.

———. 1982b. *The Impact of Oregon's Land Use Planning Program on Housing Opportunities in the Portland Metropolitan Region.* Portland: 1000 Friends.

——— and the Home Builders Association of Metropolitan Portland. 1991. *Managing Growth to Promote Affordable Housing: Revisiting Oregon's Goal 10.* Portland: 1000 Friends.

Oregon Department of Land Conservation and Development. 1991a. *Shaping Oregon's Future: The Biennial Report for 1989-91.* Salem: Department of Land Conservation and Development.

———. 1991b. *DLCD Analysis and Recommendations of the Results and Conclusions of the Farm and Forest Research Project.* Salem: Department of Land Conservation and Development.

———. 1991c. *Urban Growth Management Study, Summary Report.* Salem: Department of Land Conservation and Development.

Pease, James. 1990. "Land Use Designations in Rural Areas: An Oregon Case Study." *Journal of Soil and Water Conservation* 45, 5: 524-28.

Popper, Frank. 1981. *The Politics of Land Use Reform.* Madison: University of Wisconsin Press.

Rosenbaum, Nelson. 1976. *Land Use and the Legislatures: The Politics of State Innovation.* Washington: The Urban Institute Press.

Wall Street Journal. 1982. "The Oregon Trail." 28 October.

Whitelaw, Ed, and Ernie Niemi. 1989. "Oregon's Strategic Economic Choices." In *Oregon Policy Choices,* edited by Lluana McCann. Eugene: Bureau of Governmental Research and Service.

CHAPTER 2
Oregon's Urban Growth Boundary Policy as a Landmark Planning Tool

Arthur C. Nelson

The containment of urban sprawl has been a fundamental objective of the American planning profession since World War II. All approaches at discouraging urban sprawl have either failed or led to perverse outcomes, save one. The sole technique that has been found to be effective is the urban growth boundary (UGB), which was pioneered in Oregon. In its simplest form, the UGB places an absolute limit on urban development, which is restricted to locations within the boundary. Land outside the UGB is available for only farm, forest, or other open space uses.

Implementing UGB Policies

The UGB concept arose out of the efforts of the City of Salem and Marion and Polk counties to coordinate the management of Salem metropolitan growth. Between 1972 and 1975, the Mid-Willamette Valley Council of Governments produced several reports demonstrating the efficiencies of urban containment over urban sprawl. These efforts led to one of the first urban development stoplines adopted in the United States (Nelson 1984). This first UGB was designed to contain all the region's urban development needs to the year 2000, although subsequent analyses suggest that it will in fact do so until about the year 2020. The UGB involved approval by, and coordination with, the City of Salem, Marion and Polk counties, and water and sanitary sewerage districts. It was accompanied by extensive "down-zoning" of farm land outside the UGB to eliminate any urban development opportunities at even the lowest of densities. Land generally unsuitable for farming or other resource activities was down-zoned to 5- to 20-acre tracts and placed under restrictions to assure compatibility with nearby farm operations.

Drawing substantially from the experience of the Salem area, the LCDC wrote into the urbanization goal (Goal 14) the requirement that all incorporated cities draw UGBs. The goal reads, in part, that "to provide for an orderly and efficient transition from rural to urban land use . . . Urban Growth Boundaries shall be established to identify and separate urbanizable land from rural land." The primary function of UGBs is to manage urban growth. While growth management in other places takes the form of density constraints, development moratoria, and population caps (Scott et al. 1975), the intent of UGBs is not to limit growth but to manage its location. By restricting urban development to a well-defined, contiguous area—the size of which is based on the best available information about development trends—it is believed that growth can be accommodated without urban sprawl.

Oregon's UGB policy includes specific objectives for the preservation of prime farm land, the efficient provision of public facilities, the reduction of air, water, and land pollution, and the creation of a distinctly urban ambience. Local governments must include sufficient land within UGBs to meet the requirements for housing, industry, and commerce, recreation, open space, and all other urban land uses.

A variety of tools can be used to effect UGB policy, including tax incentives and disincentives; fee and less-than-fee acquisition of land important for land banking, public use, or open space preservation; zoning; and urban facility programming. When a UGB extends beyond a municipal boundary the county regulates the land use in coordination with the city. The UGB itself is enforced jointly by local governments and the state, while land use regulations are enforced only by local governments. This hierarchy standardizes the restrictions embodied in UGBs, while allowing variability in the management of growth within them.

Although simple in concept, the initial construction of UGBs proved difficult in practice as a result of the uncertainty about the rate and timing of urban development. Too little land could cause land price inflation; too much would not prevent urban sprawl. There were also concerns about the process of expanding, amending, or renewing UGBs. Considerable controversy arose about what should be done with land outside UGBs that was already subdivided for urban-level densities.

Although designation of UGBs was intended to be an intergovernmental effort, battles often arose between city and county governments and, in the larger metropolitan areas, between city governments. Conflicts also arose between local governments and the Department of Land Conservation and Development (DLCD), LCDC's administrative arm. Local governments frequently wanted more land inside UGBs than the state felt was justified. Most were forced to include only the amount of land needed to accommodate projected urban development to the year 2000. However, the Portland metropolitan region was allowed to have a 15.3 percent surplus and Salem 25 percent more. It appears that state participation in the land use system has resulted in less land available for urban development than would have occurred under a purely local system of land use control (Knaap and Nelson 1992). Within UGBs there are two general classes of land. "Urban" land is where most urban development already exists and where all of the immediate development needs of the urban area are accommodated. "Urbanizable land" is available for high-density development only when the supporting infrastructure is in place. Meanwhile, agricultural and low-density development is generally allowed, but only in a manner that does not preclude redevelopment at a later date.

The Public Facilities and Services goal (# 11) requires local plans "to plan and develop a timely, orderly and efficient arrangement of key public facilities and services to serve as a framework for urban and rural development." The principal aims of this goal are to direct development into urban areas, restrict development in urbanizable areas until the appropriate time, and prevent urban development in rural areas. Key facilities include water and sewer systems, and fire, public health, drainage, and recreation facilities. Facilities in rural areas can only support rural development and are mainly limited to roads, and energy and telephone lines. Water and sewer systems are discouraged in rural areas.

Unfortunately, a critical feature of UGB planning and management was left out by the LCDC when it was drafting goals in 1974. The commission decided that mandatory capital improvement programs would represent an excessive burden on local government. By 1985, however, administrative rules were revised to require them as part of all periodic plan revisions.

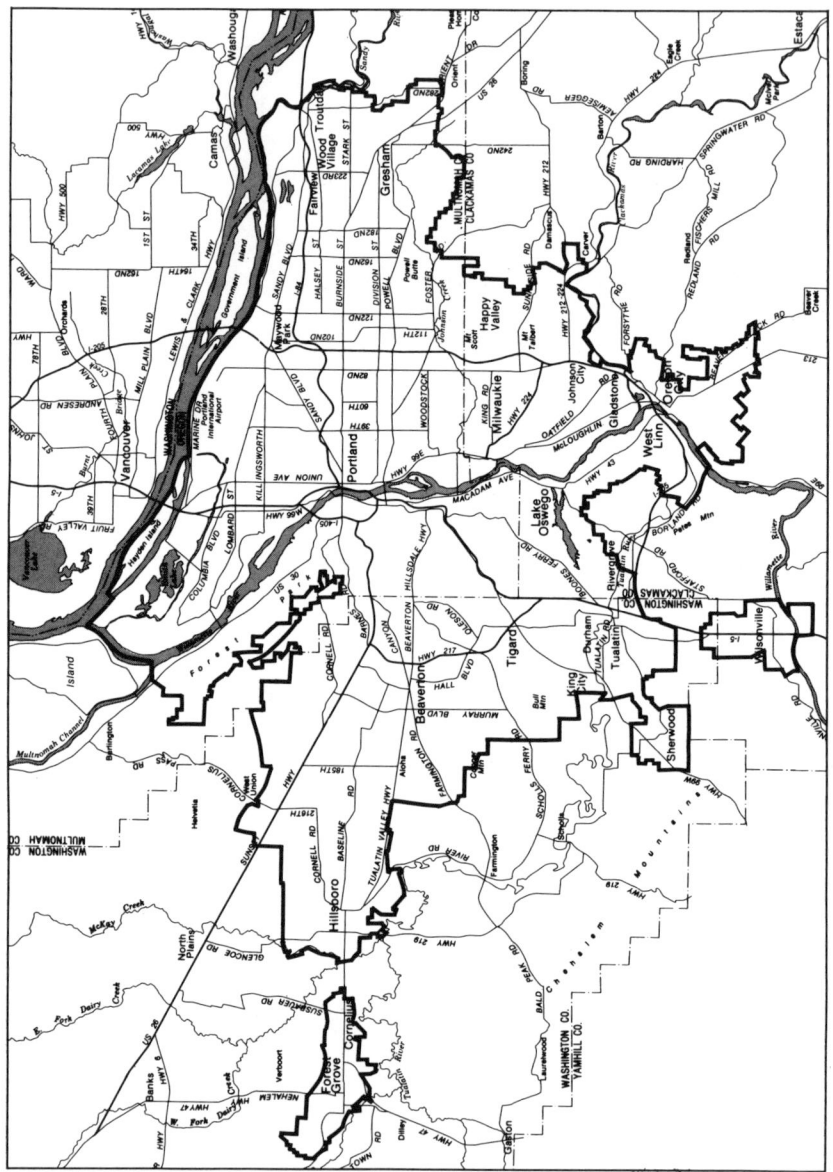

Map of Portland metropolitan area urban growth boundary courtesy of Metropolitan Service District

Theoretical and Empirical Implications of UGBs

UGBs fundamentally change the regional land market by calling into question the very assumptions of economics and tenets of property rights. Some economists argue on strictly theoretical grounds that private property owners are best able to determine the use of their own land. This arrangement would, they suggest, produce more efficient growth and development (and by implication be more effective in producing affordable housing) than would result from growth management policies. To achieve efficiency, however, the urban land market must satisfy all of seven criteria: 1) many buyers and sellers; 2) perfect information; 3) ease of entry and exit of producers within each market; 4) no transaction costs; 5) constant returns to scale in the long run; 6) buyers and sellers fully internalize the consequences of production and consumption so that nobody is made worse off by the actions of someone else; and 7) all consumers have the same tastes and preferences.

The problem is that few of these conditions can be met at any given point in time, and they can't be met simultaneously. Many inefficiencies are caused by government policies, but others are simply due to limitations of information and lack of mechanisms to ensure that benefits and costs fall on only those who cause them. Public interventions in the market aim to balance the public interest with principles of efficiency (Lee 1981), and some apparent inefficiencies are necessary to achieve public interests. For example, mortgage interest deductions against federal taxable income contribute to inefficiencies but the benefits may outweigh the costs. From a planning point of view, however, many economists and policy analysts (Bish and Nourse 1975, Ervin, et al. 1977) argue that growth management is needed largely to: 1) offset inefficient development patterns stimulated by other policies; 2) take improved (although imperfect) account of the nature of nuisances among different land uses; 3) inform buyers and sellers of overriding public interest in the environment, and desirable development patterns; 4) achieve development patterns that fulfill public policy as defined by elected representatives at all levels of government; 5) reduce negative externalities which result from interdependencies among land uses; 6) provide the optimal level of public goods; and 7) reduce the costs of providing public services.

Land economist Marion Clawson suggests a way to design a comprehensive planning system to contain urban sprawl—to counter the sprawl-inducing effects of public policy, average cost facility pricing, and market imperfections—while still accommodating the development needs of an urban area:

> If planning, zoning, and subdivision were firm— enforceable and enforced—then the area available at any one time for each kind of use would bear some relationship to the need for land for this use. That is, areas classified for different purposes could be consciously manipulated or determined in relation to market need. Sufficient area for each purpose, including enough area to provide some competition among sellers and some choice among buyers, should be zoned or classified for development, *but no more*. By careful choice of the areas concerned sprawl would be reduced, perhaps largely eliminated (Clawson 1962: 9). [Emphasis in original.]

These, then, are some of the major economic arguments for growth management generally as well as for urban containment. The trouble is that until recent years such a planning credo has been largely an article of faith. Analysis of Oregon's program sheds light on the market effects of Oregon's UGB policies. Generally speaking they paint a consistently positive picture since in most respects these policies have had desirable influences on the regional land market.

Urban Land Market Effects

Whitelaw (1980) theorized that if urban growth containment policies restricted urban development to areas inside UGBs and restricted the use of land outside UGBs to resource activities, the value of urban land would rise and the value of rural land would fall. There would be a break at the UGB in the otherwise continuously downward-sloping land value gradient. Although Beaton et al. (1977) failed to find this gap in their evaluation of the Salem UGB only a year after its implementation, Nelson (1984, 1985, 1986) did find such a gap in his evaluation of the same UGB two to four years after its implementation. Knaap (1982, 1985) found a similar gap at the Portland metropolitan UGB two years after its implementation. Since the purpose of UGB policies is to reduce if not eliminate development pressures from resource lands, these studies indicate that in this respect the policies are effective.

Facility Effects

One of the fundamental purposes of UGB policies is to make development within UGBs more efficient. This is done for the most part by making the provision of urban facilities responsive to development needs. The most effective way to achieve this is through centralized regional facility planning and administration. If sewers can be coordinated at a regional level so that the entire development community knows for certain where facilities are and when they will be extended, then development costs will be reduced. The western half of the Portland metropolitan area is served by a single sewerage agency providing the same level of service to all cities and counties it covers, but the eastern half is served by sixteen separate providers covering numerous counties and cities. Nelson and Knaap (1987) compared facility planning, land price, and development patterns between these two areas and found that the western half experienced greater benefits. Nelson (1987) went on to show that the western half also had greater fiscal capacity and lower per capita service costs. Although these studies compared two facility planning and management approaches present within the same UGB, the implications for the advantages of UGBs combined with regional facility planning are clear. Under such conditions, development costs can be cut by reducing the uncertainty of facility availability and providing facilities when and where needed. More efficient permit processing contributes as well.

Housing Effects

If there is greater demand for housing placed on a smaller supply of land resulting from UGB policies, the price of land for housing will rise. But do housing prices also rise? A house on a large lot would indeed be more expensive. On the other hand, even without UGBs constraining land supply, the natural tendency among local governments is to encourage larger lot sizes with more valuable homes and a higher tax base. The opportunities for providing smaller homes on smaller lots are correspondingly reduced.

UGB policies in Oregon are accompanied by progressive housing policies (see the analysis by Nohad Toulan in chapter 5). Local governments within the Portland metropolitan region are required by the housing rule to zone for an overall density of six to ten units per acre of vacant residential land. At least half of all residentially zoned land must allow multifamily housing or attached single-family housing. All cities and counties with populations in excess of 2,500 must permit manufactured housing on single lots. Finally, since one of the fundamental tenets of UGB policy is to accommodate rather than frustrate development, Oregon law requires development decisions within 120 days of application and further prohibits local governments from imposing arbitrary and ambiguous conditions of approval.

Farm Land Preservation Effects

In the absence of UGB policies, urban and agricultural activities compete for the same land base along the urban-rural fringe. Muth (1961) shows that where the discounted future returns to urban uses exceed those of agriculture, urban activities will outbid agricultural activities. As the supply of agricultural land declines but the demand for food increases, farming will produce greater economic benefits and will outbid urban activities. Fischel (1982) goes further by suggesting that paved-over urban land can be reclaimed for farming. Moreover, if the price of food rises above certain levels, "backyard" food production will increase.

An obvious problem with Muth's and Fischel's economic constructs, however, is that they do not consider nuisance effects that occur between urban and agricultural land uses. These include spraying (fertilizers, pesticides, herbicides), noise, hours of operation, odors, and the like. Urban residents affect farmers through trespass and petty theft of products by people and pets. In the end, it is usually urban residents who impose production-inhibiting restrictions on farming operations through local government regulations. Right-to-farm laws do not work well to protect farmers (Nelson 1990, 1992).

The result is that the value of land for farming operations is lower the closer it is to urban development. Yet the value of land for urban development is affected little by the presence of farming and can in fact be enhanced if scenic views and open space amenities exceed inconveniences. Sinclair (1967), Boal (1970), and Nelson (1992) show that agricultural land is reduced by the "shadow" effect of urban development that can extend up to 3 miles from the urban boundary (Nelson 1986).

One purpose of UGBs is to make a clear separation between urban and rural land uses to contain these shadow effects. Outside the UGB, agricultural land will be devoid of any speculative urban use value and instead will be traded solely for its farming value. However, the closer it is to the UGB the lower will be its value. Nelson found this effect in Salem (Nelson 1986) and in a slightly different context in a special exurban area of the Portland metropolitan area (Nelson 1988). On the other hand, since many urban residents prefer rural scenery to urban landscapes, the value of urban land within the UGB begins to rise as it approaches the boundary (Nelson 1986, 1988).

More telling is the effect of UGB policies on farm land production. Nelson (1992) demonstrates that the coincident effects of UGB policies and rising farming production and income relative to comparable states and the nation strongly suggest that those policies are effective in protecting farming from urban encroachments. Between 1982 and 1987, farms in the Willamette Valley increased in average size and productivity per acre. Farmers just outside the Portland UGB appear to be buying "exception" land (see definition on page 34) for more money than low-density urban households are willing to pay. Those farmers are putting land into high-value crops such as grapes (for the burgeoning Oregon wine-making industry) and berries. This would be unlikely in the absence of stable UGBs.

Urban Form Effects

Although many states have embraced elements of Oregon's statewide planning approach and some, including Florida, are arguably more assertive, none have stated the desired urban form as succinctly as Oregon. What the state is seeking is compact urban centers of variously sized cities embedded in a rural landscape devoted primarily to resource activities which itself is sprinkled with rural settlements of varying densities (see chapter 8). The effects of this urban form policy are becoming increasingly visible. Oregon gained 168,000 new residents between 1980 and 1989, most during the end of the decade as the state emerged from a forest-industry induced recession. While data on where the new residents located are unavailable, interviews by this writer indicate that more than 90 percent located inside UGBs and most of the balance in exception areas. In the Portland area alone, more than 18,000 homes and apartment units were permitted in 1989 and another 17,000 in 1990. Population in exclusive resource use areas either remained constant or actually fell, according to Census of Agriculture figures for 1987.

The theory of urban containment does not account for small-scale urban settlements in the countryside, but the planning process recognized that many rural areas outside cities and beyond UGBs were already committed to nonfarm or other nonresource uses. It was decided to allow those areas to build out rather than allowing demand for rural housing to encroach on productive farm and forest districts. More than 300,000 acres of land were set aside in the Willamette Valley for rural residential, hobby farm, ranchette, and other forms of exurban development (Gustafson et al. 1982, Nelson 1992). These areas are given "exceptions" from the strict application of statewide planning goals due to a variety of factors such as lower quality soils and the extent of existing development in the area. Restrictions must ensure that such exurban development does not adversely affect resource activities (see Knaap and Nelson 1992).

How does exurban development interact with resource activities? Nelson (1988, 1992) found property value relationships to be similar to those between urban and resource activities. Scattered islands of development adversely affect resource areas, particularly given the 3-mile shadow effect described earlier (Nelson 1986).

Management of UGB Policies

An ECO Northwest (1991) study, to which this author was an advisor on methodologicial issues, revealed several important concerns about UGB management in Oregon. Research focused on 1) the amount of post-acknowledgment residential and nonresidential development outside UGBs; and 2) the density and configuration of development immediately outside and adjacent to the UGB as constraints on future development at urban levels. The study included land inside and within 1 to 2 miles of the UGBs of the cities of Portland, Medford, Bend, and Brookings. The study period was 1988 through 1990.

Development outside UGBs

Residential development occurring outside UGBs ranged from 5 percent in Portland to 57 percent in Bend. About 17 percent of all lots created through subdividing in the Bend area occurred outside UGBs versus 3 percent for all four case studies. As of 1990, there appeared to be a capacity for about 11,250 new housing units outside the UGB in the

Portland area, 12,200 in the Bend area, about 1,500 in the Medford area, and about 200 in the Brookings area. The case studies reveal that certain counties allow much more development in farm and forest zones than others, despite the fact that they work from the same statutory base.

It is sometimes easier to build urban-like developments outside the UGB than inside. Development on exception land is not subject to the kind of technical review and development requirements imposed within the UGB. A conditional use permit for a Korean Buddhist temple was turned down in an established north Portland neighborhood, where the parishioners lived. Neighbors objected to traffic generation and an overabundance of churches. The temple was subsequently built 20 miles away on farm land just south of the Portland UGB where churches are a conditional use with few requirements.

Subdivision and development on 1- to 10-acre tracts is easier in exception areas in part because the land has been written off as neither farm land nor urban land. Low-density urban development is a consequence. These residents enjoy all the benefits of nearby urban areas without having to bear the costs. Exurban homeowners are increasingly affluent and capable of going to battle over UGB expansions to preserve their enclaves of exclusiveness. In 1990, the City of Medford was effectively blocked from expanding its UGB by owners of acreage homesites in exception areas.

UGBs were initially designed to guide urban development to 2000. Implied in the planning and acknowledgment process is that UGBs will be expanded to accommodate growth after 2000. Goal 14 does not mandate long-term land use planning consistent with the useful lives of key facilities, which can exceed fifty years. Rather, its intent is to constrain land supply in the hopes of encouraging more efficient use of urban and urbanizable land over a shorter period of time.

The trouble is that residential development in the urban fringe is resulting in a low-density residential ring around most or all of the UGB in each of the study areas. The low-density (1- to 5-acre) development makes annexations and urban service extensions more difficult. Rural areas that might have been held in reserve for future urbanization have developed in ways that will be extremely difficult to urbanize.

The twenty-year planning horizon used to establish UGBs may have contributed to the problem. Once the UGB was established, there was no requirement that urban areas plan for long-term (e.g., fifty-year) expansion needs, and no recognized obligation for counties to restrict development in areas that might be needed for long-term UGB expansion. For example, sanitary and storm drainage master planning usually considers drainage basins. Because the UGB was drawn based on a twenty-year land supply, portions of drainage basins that could have been efficiently served were placed outside UGBs and allowed to develop at rural residential densities because they were not, by definition, urbanizable. Moreover, cities and counties have not completely agreed on, or planned for, the direction of urban growth beyond the UGB. ECO Northwest (1991), Knaap and Nelson (1992) and Nelson (1984, 1986, 1988, 1990, 1992) observe that to be effective beyond the original year 2000 planning horizon, UGB policies should be modified to:

1. Require that urban areas (usually cities) establish long-term UGB expansion areas based on fifty-year public facilities needs. (This recommendation is being implemented as part of the recently adopted urban reserve rule, which will be discussed later.) Strict timelines and unambiguous standards for UGB expansion into the reserves are critical. Without them an urban reserve designation may encourage the transfer of lands from commercial farmers and foresters to those who will seek accelerated inclusion of the lands into the UGB.

2. Prohibit the placement of dwellings on land planned and zoned for exclusive farm or forest use within the urban reserve.

3. Establish a large (at least 10-acre, preferably 20-acre) minimum lot size for rural residential areas within the reserve. Require that counties notify cities of any development approvals. Require that any development or land division that is approved in the absence of urban services be conditioned upon an approved "concept" or "shadow" plan that considers the future location of urban facilities.

4. Allow for in-fill and more efficient land use in areas that are already developed at quasi-urban residential densities (one to two units per acre) and which are precluded from full urbanization in the future. Recognize that these areas are unlikely to have urban services or be annexed to a city, and give counties the authority to plan for and provide an appropriate level of services to these areas.

5. Encourage cities to include within UGBs quasi-urban areas at the urban fringe. Such a policy would encourage cities and counties to work together to provide urban services to support infill and redevelopment. An impediment to achieving this is the state's strict requirement that land

included in UGBs be justified based on the twenty-year need. That requirement encourages the city not to include quasi-urban areas in UGBs, but instead to include vacant areas that can be more readily serviced and annexed to the city. Until the annexation process is streamlined, the state could relax its strict needs requirement so that cities can include *both* needed vacant land and quasi-urban areas, thus encouraging coordinated planning for these areas. Cities must have strong conversion policies to ensure that these quasi-urban areas are not further developed without urban services.

Development inside the UGB

Because the density of residential development is below densities allowed by applicable zoning, the Bend, Brookings, and Medford UGBs may have to be expanded earlier than intended (ECO Northwest 1991 and Nelson 1990). Lots created by subdivision fell 67 percent short of allowed density inside the Bend UGB, 44 percent short inside the Brookings UGB, and 25 percent short inside the Medford UGB.

Although lots created by subdivision in Portland fell 34 percent short of allowed density, overall densities, including multiple-family development, exceeded the 6.23 units per acre assumed in justifying the size of the UGB. To help achieve affordable housing objectives within the Portland UGB, plan densities were set higher than the densities used in the UGB justification; actual densities need not meet planned densities to avoid premature UGB expansion. In Portland, as in all case study areas, however, low densities may contribute to unnecessarily high public facility costs and auto dependency.

In all case-study areas, single-family subdivisions are occurring in multiple-family residential zones. In the City of Bend, 190 subdivision lots were approved in areas zoned for multiple-family use. The densities of these single-family subdivisions were higher than the densities of subdivisions in single-family zones. However, this type of development reduces opportunities for the construction of multiple-family residences.

Commercial and industrial development in each of the four case study areas between 1985 and 1989 was concentrated inside UGBs. Less than 5 percent of the commercial and industrial developments that were constructed in the Medford and Bend areas were built outside of UGBs. Net employment changes outside UGBs in the Portland area between 1985 and 1989 were negative, implying no significant commercial or industrial development. There were a total of about 55 commercial and industrial developments created outside UGBs in the Bend, Brookings, and Medford areas.

Amounts of redevelopment and in-fill may be insufficient. In Bend and Medford, only small percentages of single-family residential development occurred in urban as opposed to urbanizable areas. Single-family development occurred primarily in subdivisions. These are easier to accommodate on large vacant parcels, which are more common in urbanizable areas. While most multiple-family units built inside the Bend and Medford UGBs were in urban areas, the number of units was far below the number of single-family units.

The effects of partitioning inside the UGB varies across case-study areas. In Medford, 56 percent of all partitions resulted in densities of four units per acre or greater; in Brookings, only 8 percent achieve those densities, due at least in part to "serial partitioning." These lots will be developed at lower than planned density, or they will continue to be re-divided to higher densities but without benefit of the coordinated planning and public services that the subdivision process is designed to provide.

Except for the Portland area, multiple-family residential development accounted for a relatively small proportion of total residential development within the primary UGB. Multiple-family units as a percent of total units were: Portland, 54 percent; Brookings, 38 percent; Bend, 21 percent; and Medford, 15 percent.

There are major differences among the case-study areas in implementing Goal 14's criteria for converting urbanizable land to fully serviced urban land. The Portland and Medford areas have developed programs that effectively limit the land divisions and low-density development inside the UGB that can occur without urban services. In the Bend and Brookings UGBs, policies that limit interim residential development are less effective, such as single-family residences are, for example, permitted without urban services on half-acre lots.

A number of policy recommendations emerge from the work by ECO Northwest (1991), Knaap and Nelson (1992) and Nelson (1986, 1992):

1. Prohibit land divisions in urbanizable areas until urban services are available or establish a large minimum lot size (10-20 acres) for areas that do not have urban services. Such measures will increase the incentive to pay for the extension of urban services necessary to support more intensive land use.

2. Establish *minimum* as well as maximum densities through zoning ordinances that specify a density range that must be achieved, rather than establishing only a density ceiling. Do not allow single-family development in multifamily zones unless a minimum density is achieved.

3. Require that any development or land division that is approved in the absence of urban services be conditioned upon an approved concept plan that considers the future location of urban facilities.

4. Prohibit serial partitioning. Require all land divisions to occur through the subdivision process and ensure that urban services are provided.

5. Require jurisdictions that allow any interim development or land divisions in urbanizable areas to have detailed public facilities plans that specify the location, source of financing, and schedule of construction for future streets, sewer, water, and storm drainage facilities.

6. Require that local zoning ordinances not allow single-family houses in urbanizable areas where land is zoned for commercial, industrial or multiple-family use.

The role of public facilities in UGB management

UGB policies will not succeed if public facilities do not accommodate development. The trouble is, who pays for those facilities and how?[1] In Oregon as elsewhere, many water and wastewater systems were financed up to 75 percent by federal grants and low-interest loans through the 1960s and 1970s. After 1993, however, federal funds were no longer available. Federal support for new roads has also been reduced.

Transportation investments in the Portland area alone will come to $3.5 billion over the next twenty years, including $1 billion for light rail expansion. If these investments are not made, especially the light rail investments, urban development would be more difficult to support. Without planning, major transportation investments, especially in the urbanizing fringe, will be followed by other major investments, such as water, wastewater treatment, schools, fire stations, and parks and recreation facilities. Such other investments may be inconsistent with Oregon's urban containment policies in certain areas and under conditions of urban sprawl. For its part, the state of Oregon has established a $100 million revolving loan fund to help urban areas provide needed infrastructure.

So how are facilities paid for? Oregon's utility statutes give great latitude to raise water and wastewater revenue through user fees, connection fees, and miscellaneous fees and surcharges. Most communities rely on state rebates of gasoline taxes to build local roads although many communities finance roads out of property taxes. For most other local facilities, the property tax has been the primary means of financing. However, Ballot Measure 5, approved in 1990, limits property taxes to $15 per $1,000 assessed valuation. Many urban areas had property tax rates approaching, and sometimes exceeding, $30 per $1,000 valuation. The new tax structure shifts financing of public schools substantially away from local governments to the state government. Both local and state governments will have less revenue with which to finance new infrastructure to accommodate development within UGBs, and without new infrastructure there will be pressure for more development in exception areas and in resource lands. In addition, one provision of Ballot Measure 5 requires a public vote on the use of tax revenue schemes, such as tax increment financing, to finance urban redevelopment. This will only complicate urban redevelopment.

Since the late 1970s, Oregon communities, especially those in rapidly growing Portland suburbs, have used systems development charges and system reservation fees to help pay for facilities. The 1989 legislature acceded to local government requests to formally enable communities to assess, collect, and spend "systems development charges." Systems reservations fees are paid by developers in advance of facility construction, such as wastewater treatment plant expansion, in order to reserve a certain capacity in that plant for their anticipated development. Troutdale used this approach to raise enough revenues to build a new wastewater treatment plant without having to depend on revenue bonds retired, at least initially, on a user base that was less than one-tenth of the eventual number of users.

Since 1986, however, the post-acknowledgment period of planning now requires local governments to submit plan revisions and updates every two to five years. Under this procedure, local governments must prepare and implement capital improvement plans that show how facilities will be financed. The Oregon legislature created a "special public works fund" in 1991. The fund uses the borrowing capacity and favorable bond rating of the state to make financing of local government infrastructure more affordable. The state pays all costs of issue. Local governments essentially borrow money from the state for less than they would pay to borrow elsewhere. Each year, local governments submit loan proposals, which the state then packages into a large issue. The fund is authorized for $100 million, but by 1993 it had only $15 million outstanding. Because of Ballot Measure 5, many local governments are reluctant to increase their debt, at least until the implications of the measure for local fiscal structure become better known.

The role of facility financing in achieving Oregon's urban containment objective has been understated by the LCDC and DLCD, if not the legislature itself. An analysis by the Center for Urban Studies, Portland State University, and Regional Financial Advisors, Inc. for DLCD (1990) revealed several limitations in the ability of Oregon's local governments to adequately finance needed infrastructure such as:

1. Funding from local and state sources accounts for only about half of all anticipated infrastructure needs of local government.

2. State aid finances only about one-fifth of local government needs for roads, water, and sewer facilities to accommodate urban development.

3. The state may not provide sufficient support for the financing of schools, parks, open spaces, libraries, and police and fire facilities.

4. State loan opportunities are very nearly as expensive as market sources of loans such as bonds, in contrast to many other states where low-interest loans are made to local government for infrastructure expansion.

5. Local governments do not make maximum use of user fees, systems development charges, or special assessment districts to finance infrastructure expansion. Part of the problem is statutory limits but local governments are also reluctant to increase debt financing burdens to accommodate new development.

6. Ballot Measure 5 will limit local financing options and exhaust state resources.

Existing local public facilities plans are probably not up to the tasks that long-run growth management wants them to perform. Based on ECO Northwest's (1991) review of those plans, they concluded that 1) the state does not have a consistent standard for the review of public facilities plans; 2) responsibility for determining needed public facilities projects (and estimating their costs and timing) is sometimes unclear; and 3) acknowledged public facilities plans have not been prepared at a sufficient level of detail or accuracy to make useful cost comparisons. The City of Bend looked only at city sewer and water facilities in developing its public facility plan, and Bend's most expensive transportation project—the Bend Bypass—is not identified in its capital facilities plan.

In sum, financing new facilities needed to accommodate new development has become more problematic in recent years rather than more predictable or efficient. As a result, facilities may not be in place or programmed concurrent with new development. Some urban development may be discouraged from locating in urban areas. Other development will proceed but only by congesting or otherwise degrading existing facilities. Certain local governments will expand the use of systems development charges but others may deny development permits arguing insufficient existing facilities and services.

Of additional interest is the role of facilities in directing development into urban areas. This may seem an obvious proposition but the issue is complex. For example, the majority of state highway expenditures in the Bend, Brookings, and Medford case-study areas occur outside of UGBs and may work against state land use policies intended to concentrate urban growth inside UGBs. In the case study's three less-urbanized counties (Curry, Deschutes, and Jackson), rural areas accounted for about 85 percent of state highway expenditures, compared to only 24 percent in the three Portland-area counties. The Oregon Department of Transportation's mission of connecting urban areas requires expenditures on highways in rural areas. Such expenditures enhance the attractiveness of rural housing opportunities by aiding access to them.

What Are the Future Challenges?

In Oregon, future challenges will require redoubling of the state's commitment to preserving resource lands and forcing compact urban development. The whole planning program was geared in 1974 to plan for the year 2000. Indeed, all UGBs were designed to accommodate urban development needs to that year. There has been a lot of speculation as to what happens after 2000.

Tim Ramis, a Portland land use attorney, suggests that UGBs and most urban containment policies may "sunset" in the year 2000. He does not think this is a likely outcome, but it will not be litigated until at least January 1, 2000. Gerrit Knaap of the University of Illinois suggests that UGBs will remain in place and for all intents and purposes will not be extended.

Opposition by residents of exception areas could contribute to a fixed UGB.

One other idea advanced by this author is for selected UGB expansion into urbanizable exception lands and creation of highly dense satellite towns. The satellite towns would be strategically located outside major urban areas; created out of existing towns where political, economic, and social infrastructures exist to guide planning and development; linked to major urban areas via transportation corridors including light rail if economically feasible; and put into place through a combination of new statutory authorities enabling private sector initiative in creating the satellite towns through redevelopment.

In part because of concerns over managing exception areas adjacent to and near UGBs, and managing UGB expansion into those areas, the LCDC adopted its Urban Reserve rule in 1992. The rule must be implemented by seven cities: the Portland metropolitan area, Newberg, Hood River, Sandy, Grants Pass, Brookings, and Medford. These areas were selected for mandatory application of the rule because of their population growth, population size, and amount of development in nearby exception land. Other cities may also implement the rule but they are not required to do so.

By 1995, these seven urban areas must 1) temporarily stop up-zonings in exception areas near UGBs; and 2) establish the extent to which the UGB may be extended into certain exception areas over time. The urban reserves will include some farm and forest lands, but only to the extent that development patterns, soil conditions, and related factors indicate that such lands are necessary for conversion to urban uses.

In effect, the urban reserve rule indicates that LCDC has embraced the third approach by creating the mechanism by which UGBs may be expanded into urbanizable exception areas, and by which some outlying communities may become larger satellite towns. It is interesting to note that the LCDC requires application of the urban reserve rule in Newberg, Hood River, and Sandy, all of which are within commuting range of the Portland metropolitan area.

It is possible that the urban reserve rule will not be effective in managing development of exception lands and allowing for timely expansion of the UGB. The problem is that the rule tries to correct the original mistake of not anticipating development of exception areas near UGBs. But the horses are out of the barn; many exception areas within the urban reserve are substantially developed and occupied by people who will fight UGB expansion.

In the meantime, pressure to expand the UGB is not yet evident. Developers have little problem finding adequate parcels for reasonable prices. But the time will come when the large easy-to-develop sites are gone. That is when developers will test the UGB policies. It is expected that once sites are exhausted out to the UGB, developers will hunt for in-fill and redevelopment sites closer in. Ethan Seltzer of the Metropolitan Services District calls this the "back wave" (Nelson 1990). However, no one knows how developers and the market will actually respond.

While Oregon policy makers argue that strict farm land preservation policies were aimed at sustaining the commercial agriculture industry, few analysts believe that was the only motive. A major concern was simply to provide a scenic backdrop of open spaces around urban areas. Given this perspective, some urban areas may consider arrangements to buy farm land development rights. Indeed, Rena Cusma, executive director of the Metropolitan Service District, has directed her staff to begin investigations into development rights acquisition around the Portland area UGB.

An "Ultimate" Urban Form?

The Oregon Land Use Act of 1973 aims to contain urban sprawl within UGBs to the year 2000. That year is fast approaching. At a minimum, the following questions have yet to be addressed:

1. Will resource lands continue to be preserved from any kind of development?

2 Where will the needs of urban development be accommodated in some way during the next century, if not within the present limits of UGBs?

3. Should there be a reassessment of the hierarchy of cities and urban places to identify those places that are ascending and should continue to ascend in their geo-economic prominence? Should some places be allowed or encouraged to rise in regional economic prominence, while other places are maintained at present or lower levels of relative prominence? The urban reserve rule begs this question.

4. Should there be clear consideration of the potential need for satellite towns linked by a variety of transportation and communication technologies?

5. What should be the role of government in facilitating greater urban in-fill, redevelopment, and intensification of land uses? Should government intervene in the land market to facilitate urban land conversion processes if planning policy is frustrating these processes?

6. Should infrastructure resources be more consciously diverted to urban areas to further facilitate compact urban development?

7. Should means be found to permanently acquire the development rights of certain resource lands to form permanent greenbelts in such size, shape, and proximity to markets as to sustain a critical mass of resource activity?

8. Should government become more active in providing affordable housing?

9. Should the state embark on statewide user-fee schemes to finance state-funded facilities particularly highways, and impose marginal cost pricing on users as an explicit way in which to reward people who choose efficient use public facilities and services?

10. To what extent should the state rescind policy-driven utility, facility, and household subsidies that militate against compact urban development in favor of urban sprawl?

The logical extension of Oregon's statewide land use planning program is the achievement of an ultimate urban form not only for the highly populated Willamette Valley but for the entire state. The outline of this ultimate urban form is already in place. It is composed of urban areas contained within UGBs; preservation of resource lands for nonurban uses outside UGBs; the formal establishment of greenway corridors and the informal establishment of greenbelts (working farms) around urban areas; greater development occurring inside UGBs than outside; greater regional coordination of planning and administration in the major urban centers; and implementation of the urban reserve rule, which will result in selected expansion of UGBs into exception areas and increased development of some satellite communities in and near the Portland metropolitan area. It is supplemented by a reviving agricultural economy combined with increasing political awareness of UGBs and policies to preserve resource lands among the citizenry. The challenge facing Oregon now is how to properly recognize the urban form it has created through UGB policies and its implications, in what manner it should be reassessed, and how best to consciously facilitate that urban form. Now is the time for Oregon planning institutions to think ahead to the twenty-*second* century.

Notes

1. In an economic sense those who pay for infrastructure can be different from those who actually write the checks. For example, developers may write the checks for systems development charges and connection fees, but the home buyer actually pays the bill. In some situations, taxpayers may pay the bill.

References

Beaton, C. Russell, James S. Hanson, and Thomas H. Hibbard. 1977. *The Salem Area Urban Growth Boundary*. Salem: Mid-Willamette Valley Council of Governments.

Bish, Robert, and Hugh Nourse. 1975. *Urban Economics and Policy Analysis*. New York: McGraw-Hill

Boal, Frederick W. 1970. "Urban Growth and Land Value Patterns." *Professional Geographer* 22, 2: 79-82.

Center for Urban Studies and Regional Financial Advisors, Inc. 1990. *Local Government Infrastructure Funding in Oregon*. Salem: Department of Land Conservation and Development.

Clawson, Marion. 1962. "Urban Sprawl and Speculation in Suburban Land." *Land Economics* 38, 2: 99-111.

ECO Northwest. 1991. *Evaluation of Urban Growth Boundary Policies*. Salem: Department of Land Conservation and Development.

Ervin, David E., James Fitch, R, Kenneth Godwin, W. Bruce Shepard, and Herbert Stoevener. 1977. *Land Use Control: The Economic and Political Effects*. Cambridge: Ballinger Publishing Co.

Fischel, William. 1982. "The Urbanization of Agricultural Land." *Land Economics* 58, 2: 236-259.

Gustafson, Greg C., Thomas L. Daniels, and Rosalyn P. Shirack. 1982. "The Oregon Land Use Act." *Journal of the American Planning Association* 48, 3: 365-73.

Knaap, Gerrit J. 1982. *The Price Effects of an Urban Growth Boundary*. Ph.D. Dissertation. University of Oregon: Eugene.

———. 1985. "The Price Effects of an Urban Growth Boundary in Metropolitan Portland, Oregon." *Land Economics* 61, 1: 26-35.

———, and Arthur C. Nelson. 1992. *The Regulated Landscape: Lessons on State Land Use Planning from Oregon*. Cambridge: Lincoln Institute of Land Policy.

Lee, Douglass B. 1981. "Land Use Planning as a Response to Market Failure." In *The Land Use Policy Debate in the United States*, edited by Judith Innes deNeufville. New York: Plenum Press.

Muth, Richard F. 1961. "Economic Change and Rural-Urban Land Conversions." *Econometrica* 29, 1: 1-23.

Nelson, Arthur C. 1984. *Evaluating Urban Containment Programs*. Portland: Center for Urban Studies, Portland State University.

———. 1985. "Demand, Segmentation, and Timing Effects of an Urban Containment Program on Urban Fringe Land Values." *Urban Studies* 22, 5: 439-443.

———. 1986. "Using Land Markets to Evaluate Urban Containment Programs." *Journal of the American Planning Association* 52, 2: 156-171.

———. 1987. "The Effect of a Regional Sewer Service on Land Values, Growth Patterns, and Regional Fiscal Structure within a Metropolitan Area." *Urban Resources* 4, 2: 15-18, 58-59.

———. 1988. "An Empirical Note on how Regional Urban Containment Policy Influences Interaction between Greenbelt and Exurban Land Markets." *Journal of the American Planning Association* 54, 2: 178-184.

———. 1990. "Economic Critique of Prime Farmland Preservation Policies in the United States."*Journal of Rural Studies* 6, 2: 110-42.

———. 1992. "Preserving Prime Farmland in the Face of Urban Development." *Journal of the American Planning Association* 58, 4: 467-488.

———, and Gerrit Knaap. 1987. "A Theoretical and Empirical Argument for Centralized Regional Sewer Planning." *Journal of the American Planning Association* 53, 4: 479-86.

Scott, Randall W., with David Brower and Dallas Miner. 1975. *Management and Control of Growth*, three volumes. Washington: Urban Land Institute.

Sinclair, Robert. 1967. "Von Thuenen and Urban Sprawl." *Annals of the Association of American Geographers* 57, 1: 72-87.

Whitelaw, W. Ed. 1980. "Measuring the Effects of Public Policies on the Price of Urban Land." In J. Thomas Black and James E. Hoben, eds., *Urban Land Markets*. Washington: Urban Land Institute.

CHAPTER 3
The Legal Evolution of the Oregon Planning System

Edward J. Sullivan

The Oregon planning system involves a "federal" approach to land use planning and regulation. A state agency, the Land Conservation and Development Commission (LCDC), was created by statute in 1973 to adopt mandatory goals for local governments (i.e., cities and counties) to incorporate into their comprehensive plans. The commission also has the power to adopt administrative rules to elaborate upon, or interpret, the goals. LCDC is staffed by the Department of Land Conservation and Development (DLCD), which implements commission policy.

The real work of planning, however, goes on at the local level with some state funding. Cities and counties are required to adopt comprehensive plans which provide articulable standards for development to be implemented through zoning, land division, and other regulations. Local governments must coordinate their plans and activities with the work undertaken by special districts (which provide school, water, and other services) and state agencies.

In addition, the state has divested courts of jurisdiction over most land use matters and given that jurisdiction to the Land Use Board of Appeals (LUBA), which, with some exceptions, reviews state agency, special district, or local government land use decisions. Review of LUBA decisions is undertaken at the appellate court level.

As this chapter will illustrate in detail, the system is predicated upon a proactive state agency, LCDC, setting statewide standards (or "goals") and enforcing their incorporation in local plans and regulations by cities and counties. The system is heavily influenced by private interest groups and individuals who are concerned with application and enforcement of state policy and who seek to move public agencies in a certain direction. LCDC, LUBA, and the courts provide several forums for doing so.

A number of characteristics set Oregon's program apart from those of other states. These include:

1. A heavy emphasis on procedure in quasi-judicial decision making, including notice, opportunity to be heard, and a reasoned explanation for decisions made in the light of the facts found and the law to be applied.

2. A requirement that land use decisions not be *ad hoc*, but based upon previously determined policy set forth in a city or county comprehensive plan.

3. A requirement that local comprehensive plans be based upon state policy set forth in the statewide planning goals and be the result of citizen participation.

4. A policy emphasis on conservation of agricultural and forest lands and natural resources.

5. A policy that requires cities and counties to agree upon urban growth boundaries, which separate "urban lands" (which include cities) from "rural lands" (areas with sparse settlement where urban growth is not expected). Housing needs must then be identified and a means formulated to provide for sufficient density and types of dwellings to meet that need. Infrastructure needs within urban areas must also be determined.

6. Means by which plans are reviewed and found to be in compliance with the state goals. Plans and implementing regulations, and any amendments, are reviewed by the state for continued compliance with the goals. Enforcement action is available.

7. Finally, a state agency (LUBA) is provided to review those public agency land use decisions upon request by other public agencies, interest groups, and citizens.

The purpose of this chapter is to give the reader the legal framework of the Oregon program. Much is omitted for reasons of space. Additional information is available from DLCD, the Oregon State Bar, and 1000 Friends of Oregon (a public interest group which often participates in proceedings before LCDC and LUBA).

The establishment of Oregon's planning system in 1973 may well have been a historical accident. It came at a time when Tom McCall, a very popular governor, found a receptive audience for his warnings that, without planning, the state would lose its most fertile farm and forest lands, spoil its coastlines, and ignore its housing needs. It followed an unsuccessful legislative effort in 1969 to require the state's cities and counties to adopt comprehensive plans and zoning regulations. The time was also ripe for the Oregon Supreme Court to consider the nature of planning and its relationship to regulation of the use of land. A happy coincidence of concern over the natural and human environment, a commonly held belief that planning and regulation could avoid future problems, and an enlightened judiciary all occurred at the same time.

One of the more significant features of the Oregon planning system is that zoning and other forms of land use regulation are subordinate to the comprehensive plan. This principle was actually established in case law before the program was fully implemented. In *Fasano v. Board of County Commissioners of Washington County*,[1] the Oregon Supreme Court determined that the state's zoning enabling legislation required counties to have comprehensive plans and to carry out those plans in their zoning, subdivision, and other land use regulations. That court came to the same conclusion with regard to cities in *Baker v. City of Milwaukie*.[2] In that case, the court likened the comprehensive plan to a constitution for land use decision making.[3]

As a result of the *Fasano* and *Baker* decisions, citizens may expect public policy to be articulated in the comprehensive plan and carried out in zoning and other implementing regulations. In addition, the comprehensive plan provides a policy foundation upon which rezoning and other land use decisions are to be based rather than having policy decided on a piecemeal, *ad hoc* basis as under the majority interpretation of Section 3 of the federal Standard Zoning Enabling Act (Sullivan and Kressel 1975).

Because the comprehensive plan is the basis for zoning and other land use regulatory actions, Oregon courts do not utilize a substantive due process analysis when those actions are challenged. About fifteen states have followed the *Fasano* rationale. As noted elsewhere (Sullivan 1990), "[m]any state courts, however, were skeptical of that rationale, perhaps because of the longstanding identification of the comprehensive plan under the Standard Act with the zoning map, and perhaps because there was no requirement in most states that there be a comprehensive plan."

State Goals and Local Plans

Goal formulation and amendment

LCDC has the responsibility of establishing state policy through the adoption or amendment of goals, or planning standards, which apply to state agencies, special districts, and local governments through local government plans and implementing regulations.

Though there are mandatory considerations in the adoption and amendment of goals, such as existing state agency, local government, and special district plans, and a list of additional considerations found in ORS 197.230 (b), in practice these considerations have rarely affected the formulation or amendment of the existing nineteen statewide planning goals. LCDC has used its own compass in determining which goals to adopt or amend.

The commission has been required since 1977 (Chapter 664, Oregon Laws 1977) to make a finding of need for the adoption or amendment of a goal. In practice, that finding has been relatively easy to make. Under that same legislation, LCDC must also design goals to provide a reasonable degree of flexibility in their application. LCDC is prohibited from making its goals "specific land management regulations." The goals, however, have provided the basis for such regulation, particularly in the Willamette River Greenway and coastal areas.

A detailed process for goal adoption and amendment requires DLCD to hold at least ten hearings throughout the state. After the hearings, the draft goal or amendment must be submitted for comments from the Citizen Involvement Advisory Committee, the Local Officials Advisory Committee, and the Joint Legislative Committee on Land Use. The commission itself must hold at least one public hearing. If the draft is adopted, the commission must establish a schedule for an effective date for the new or amended goal, which must be at least one year from adoption, unless compelling circumstances dictate otherwise.

DLCD is responsible for notifying local governments of the adoption of any new or amended goal, administrative rule, and land use legislation. In each case, the local government must amend its plans and regulations during the post-acknowledgment and periodic review processes. If a local government fails to take action, that failure may be the basis for enforcement action.

Acknowledgment

The process by which the goals are applied through local government plans and regulations involves both acknowledgment and enforcement. Acknowledgment is the formal recognition by LCDC that local plans and regulations, read together, meet the goals.

Initially, the legislature and the planning community had failed to understand how significant the decision to recognize compliance with the goals would become. With the adoption of statewide planning goals in 1974 and 1975, local government decisions were required to meet both the goals and local plan and regulatory provisions. An important land use decision could be delayed or defeated for failure to demonstrate compliance with the goals. Moreover, a body of case law grew up around the goals. Local governments and the development community had a significant incentive to secure acknowledgment so that only local plans and regulatory provisions, which, at least in theory, met the goals, would control local decision making. Although acknowledgment was not part of the original Senate Bill 100, it was added in 1977 to provide a point at which formal recognition could be undertaken and that decision reviewed by dissatisfied participants.

The 1977 legislative session that formalized acknowledgment also provided for administrative and judicial review of the acknowledgment decision. Acknowledgment is requested by cities and counties and, in one case, by the Metropolitan Service District. The acknowledgment request is submitted to the DLCD director, who then prepares a detailed report analyzing whether the plan and regulations meet each of the applicable planning goals (some goals, such as the coastal goals, are not applicable to all jurisdictions).

ORS Ch.197 specifies in detail the process for obtaining oral and written comments, establishes which persons or governmental agencies have the right to contest acknowledgment requests, and provides a procedure for those submitting comments to take exception to the director's draft report. The commission reviews the plan and implementing regulations, the comments of objectors and others, the draft staff report, the exceptions, and any supplemental response by the director.

LCDC might respond in three ways to an acknowledgment request:

1. The request may be granted, subject to appeal to the Court of Appeals. This means that the local plan and implementing regulations are

the sole criteria for state agency, special district or local government land use decisions.

2. Acknowledgment may be denied because the plan and implementing regulations are at variance with the goals. This rarely occurs; if it did, enforcement action could be taken.

3. The most frequent action is a continuance (usually from 30 to 180 days) to give the local government an opportunity to cure specified deficiencies in its plans and regulations as detailed in a compliance schedule. Those aspects of the plan found to be in compliance are appealable to the Court of Appeals.

At this time, all 241 cities and 36 counties have been acknowledged, so that the initial acknowledgment process is no longer relevant. If a new city were created on rural land and sought to urbanize, this process might again be used; however, most new cities will be created on urban land and will already have an urban planning background from which to work. The attention of local governments in Oregon is now more focused upon the periodic review process, discussed below, in which plans are reviewed on a four- to ten-year basis to determine continued compliance with the statewide planning goals.

Rule making

LCDC has the power to adopt administrative rules to fulfill its responsibilities. These include normal state agency rules required under the Oregon Administrative Procedures Act such as adoption of rules, contested case procedures, and public records. There are also matters relating to express statutory grants, former statutes which allowed the commission to hear cases filed with it relating to the goals, acknowledgment procedures, application of goals to incorporation of new cities, adoption of the goals themselves, post-acknowledgment and periodic review processes, and state agency coordination. Of greatest importance, however, are LCDC's rules dealing with the interpretation of the statewide planning goals.

Previously, DLCD assembled policy papers that interpreted the goals, but failed to adopt these papers as rules. The legislature prohibited this practice in 1981, confirming the predilection toward rule making. Because the goals frequently use loose language, rules often interpret and clarify the meaning of the terms. Specific rules have been developed covering agricultural lands, forest lands, housing in the Portland Metropolitan area, housing, citizen involvement, public facilities planning, natural resources, classifying Oregon estuaries, the Willamette River Greenway, and urban reserves.

Enforcement

The process of enforcing the implementation of the statewide goals has become a complicated task since it was first authorized by the legislature in 1977. Under the current system, persons other than the DLCD or LCDC may request an enforcement order by presenting written reasons to the affected local government and requesting relevant changes. The requestor and local government may enter into mediation and the department may join if the other parties so request. If the requestor is not satisfied with the local response, he or she may present a petition to the commission.

The commission is authorized to order a local government, special district, or state agency to bring its plan, regulations, or decisions into compliance with the goals or acknowledged plans and regulations. In determining that an order should be issued, LCDC must state the nature of noncompliance and necessary corrective actions. This final order is reviewable by the Court of Appeals. The order can include development limitations and the withholding of state shared revenues. The commission may institute further judicial action to enforce its order.

Post-acknowledgment review

It soon became apparent that much would be lost if plans and regulations were acknowledged but the state no longer had a role in participating in amendments to these plans and regulations. In 1981, the legislature approved detailed procedures to enable LCDC to review these amendments. Local governments must send notices of proposed amendments to the director at least 45 days before the final hearing on the amendments. No notice is required if the local government believes the goals do not apply to the proposal or there are emergency circumstances requiring expedited review. In either case, the local government must submit a copy of the adopted amendment to the director and any person may appeal the amendment to LUBA.

DLCD is required to give notice of plan and regulatory amendments to those who have so requested. The department can participate locally in the amendment process by notifying the local government of any concerns and recommendations it has. However, DLCD may choose not to participate in local proceedings. Upon adoption, the local government must send a copy of the amendment along with findings justifying approval to the director. Anyone who participated locally in the adoption of the amendment may make an appeal to LUBA. The director and any other person who did not participate may appeal the amendment if the adopted version differs from that originally submitted to the director.

Unless appealed, or if affirmed on appeal, the amendment is deemed acknowledged. The director or LUBA, as appropriate, are authorized to issue a certificate of acknowledgment upon request of any person.

Periodic review

In another major program adjustment similar to the acknowledgment process, the Oregon legislature in 1981 instituted a formal process for periodically reviewing acknowledged plans and implementing regulations to assure their continued compliance with the goals. This process was substantially revised by the 1991 legislature.

Local governments are required to review their plans and regulations in accordance with a schedule adopted by LCDC, usually between four and ten years after acknowledgment or the last periodic review. LCDC can develop a schedule to allow regional coordination of periodic review.

The department begins the process by notifying the local government that it is scheduled for periodic review and outlining review requirements along with the information which must be submitted. The local government then reviews its citizen participation program, plan, and regulations.

On the basis of self-examination, the local government may determine that no work program is necessary, either because the plan and regulations meet the goals or because the plan and regulations have been amended through the post-acknowledgment process. Otherwise, the local government must develop a work program for meeting the periodic review criteria with a completion date for each task. The criteria are:

1. There has been a substantial change in circumstances in the findings or assumptions upon which the comprehensive plan or land use

regulations were based, so that they are out of compliance with statewide planning goals.

2. Implementation decisions and their effects, are inconsistent with the goals.

3. There are issues of regional or statewide significance, intergovernmental coordination of state agency plans, or programs affecting land use which must be addressed in order to bring comprehensive plans and land use regulations into compliance.

The local government transmits its assessment to the director, who determines whether the periodic review criteria have been met and may also coordinate work programs. The director may approve the evaluation and work program or the local determination that no work program is necessary, reject that evaluation and work program and suggest resubmittal by a specified date, or refer the evaluation and work program to LCDC.

The director's decision may be appealed to the commission under its administrative rules. The commission's decision is appealable to the Court of Appeals. The work program stage is important, for the tasks normally may not be revisited later if the local government or any other party wishes to challenge that stage. The commission may modify an approved work program under limited circumstances. Once adopted, the work program is implemented through adoption of plan or regulatory provisions and submission at that point to the director. Persons objecting to the conformity of those tasks to the work program and the goals may appeal the same to the director, the commission and the appellate courts.

State Goals and Intergovernment Coordination

State agency coordination

One of the most difficult parts of the state's planning program deals with the relationship between acknowledged local government plans and regulations on the one hand and state agency programs and rules on the other. The structure and theory of the Oregon land use program is that local governments are the primary units for land use planning activity. LCDC may adopt goals and review local plans and regulations for conformity, but it is the local government which is the medium for expression of that policy. This means that state programs are at least theoretically subject to coordination and approval by local governments. Relatively few issues have arisen in this area, but great potential for conflict remains.

With the express exemption of the Forest Practices Act or other programs exempted by another statute, state agencies "shall carry out their planning duties, powers and responsibilities and take actions that are authorized by law with respect to programs affecting land use" in compliance with the statewide planning goals *and* acknowledged local plans and regulations. The nature of this dual obligation is virtually unexplored in case law.

Something akin to the process of acknowledgment occurs with regard to state agencies—certification of state agency rules and programs. Upon request of the department, state agencies must submit: 1) their rules and summaries of programs affecting land use; 2) a program for coordination to assure compliance with state goals *and* compatibility with acknowledged city and county comprehensive plans;[4] 3) a program for coordination with federal agencies, other state agencies, local governments and special districts; and 4) a program for cooperation with and technical assistance to local governments.

The director reviews the submittals and forwards findings as to compliance and compatibility to LCDC, which must either certify the agency's coordination program or determine that it is insufficient. Until an agency is certified, it must make findings when adopting or amending its programs "as to the applicability and application of the goals or acknowledged comprehensive plans."

LCDC must also adopt rules to assure that state agency permits are issued in compliance with the goals and consistent with acknowledged comprehensive plans and regulations. The rules are required to state the extent to which state agencies may rely upon local determinations of compatibility. If the plan and regulations are not acknowledged, the state agency must be supported by independent goal compliance findings.

State agency programs, rules, or actions are not deemed compatible with acknowledged plans if the action is not allowed under that plan. The agency may apply its own statutes and rules to condition or limit a use to make that use conform to the acknowledged comprehensive plan. A state agency is exempted from acting compatibly with local acknowledged plans if its plan or program relating to land use was not in existence at the time of acknowledgment of the local plan and regulations and the agency demonstrates that: 1) the plan or program is mandated by state or federal law;

2) the plan or program is consistent with the goals; 3) the plan or program has objectives which cannot be achieved in a manner consistent with the local plan and regulations; and 4) the agency has complied with its certified state agency coordination program. This exemption is yet to be tested.

There is actually very little experience with state agency coordination, as the courts weigh the delegation of legislative authority to agencies other than LCDC against the state agency coordination statute and are reluctant to equate coordination with operational control over state government. The next two decades of state planning will no doubt resolve these difficult issues.

Local government coordination

Each county is responsible for coordinating all planning activities affecting land use within the county, including the activities of the county, cities, special districts, and state agencies to assure an integrated comprehensive plan for the entire county. By statute, "[a] plan is 'coordinated' when the needs of all levels of governments, semipublic and private agencies and the citizens of Oregon have been considered and accommodated as much as possible."[5] While this definition provides little specific guidance, two hallmarks have been identified for a properly coordinated plan:

1. The makers of the plan engaged in an exchange of information between the planning jurisdiction and affected government units, or at least invited such exchange; and

2. The jurisdiction used this information to balance the needs of all governmental units as well as the needs of citizens in the plan formulation or revision.[6]

Counties and cities may elect to form a regional planning authority to exercise the coordination responsibility granted to counties.[7]

By state mandate that dates back to SB 100, the general authority to plan and regulate land use within the Portland metropolitan area, which encompasses most of Multnomah, Clackamas, and Washington counties, is exercised by the Metropolitan Service District (Metro).[8] This authority includes the adoption and enforcement of regional land use planning goals and objectives, the preparation and enforcement of functional plans (air and water quality, transportation, etc.), designation of areas and activities of significant regional impact, review of local comprehensive plans, and the coordination of planning within the region, including the review and coordination functions which would otherwise be required of the three counties within the agency's territorial planning jurisdiction.

Special districts provide within unincorporated areas public services typically provided by cities. They do not have an obligation to prepare comprehensive plans, but must exercise their planning responsibilities in accordance with LCDC goals. In 1977, the legislative assembly provided a mechanism for facilitating coordination between special districts and counties. Each special district must enter into a cooperative agreement with the county in which it is located.[9]

The Land Use Board of Appeals

The Land Use Board of Appeals (LUBA) was first created in 1979 as a means of consolidating review of land use decisions of local governments, state agencies, and special districts. The legislature wished to remove review of local land use decisions from the circuit courts of the state, where review was lengthy, costly, and undertaken by judges who rarely saw land use cases. Circuit courts either took cases by way of declaratory or injunctive relief, which normally required a trial and dealt with matters of legislative policy, or by the ancient Writ of Review, for quasi-judicial decisions of local government in which policy was applied to a particular case. There was no statute of limitations for declaratory or injunctive relief, while there was a sixty-day period for use of the Writ of Review. These methods were very costly, as a practical matter required the use of lawyers, and often resulted in decisions which contradicted each other on legal points so that a further appeal was required. The major reason for the change is set forth in the legislative policy for review of land use decisions, i.e. that time is of the essence in undertaking such review.

The board consists of three members appointed by the governor and confirmed by the Senate for four-year terms. Board members must be attorneys admitted in Oregon. The board's main offices are in Salem, but it may hear cases in any part of the state. While each referee may sit independently on a case, the practice is to hear cases as a full board. The chief referee then assigns a board member to write the opinion.

LUBA was granted power to review both legislative and quasi-judicial land use decisions of a local government, state agency, or special district. The term "land use decision" is critical to an understanding of

LUBA's jurisdiction, for the board has exclusive jurisdiction over such decisions, except as provided by statute. ORS 197.015(10) defines such a decision as one involving a local government or special district adoption, amendment, or application of the statewide planning goals, a comprehensive plan provision, or a land use regulation, or a state agency decision where the agency is required to apply the goals. As a practical matter, Oregon appellate courts have given LUBA a wide jurisdictional latitude, as they find a legislative direction to do so. As a result, issues as disparate as street improvements, urban renewal decisions, and minor land partitions have been found to be land use decisions.

The grant of jurisdiction to LUBA does not affect other grants of land use regulatory powers, such as that to LCDC to adopt and administer the statewide planning goals and to the Court of Appeals to undertake judicial review. For state agencies, however, the Court of Appeals has exclusive review authority unless the land use decision is an "order in other than a contested case," which are heard by LUBA instead of the circuit court. Thus, most state agency decisions will be reviewed by the Court of Appeals, rather than LUBA.

There are other limitations to LUBA's review authority:

1. LUBA has no authority to review forest practices rules, programs, decisions, determinations, or activities. This exclusion reflects the political power of the timber industry in Oregon. These matters rest with the circuit courts or the Court of Appeals.

2. Circuit courts retain jurisdiction to deal with enforcement of local land use regulations and LUBA orders. However, many local governments have found it cheaper and easier to have internal enforcement mechanisms requiring a participant to take an appeal to LUBA within 21 days of the local enforcement decision as a means of containing enforcement costs. Thus, the use of the trial court system in Oregon land use appeals is minimal.

3. LUBA also lacks authority to review ministerial acts, i.e. those involving determinations in which no discretion is exercised, such as the issuance of building permits, or where only factual determinations (e.g. the calculation of a setback) are involved.

4. LUBA does not have authority in those areas delegated to LCDC, including acknowledgment or periodic review.

In all cases, the challenged decision must have been given after exhaustion of all remedies available by right.

LUBA normally confines its review to the record before the local government. In rare cases where matters outside the record are cause for reversal or remand, LUBA may hold evidentiary readings. The board must render its final order within 77 days of the date the record is settled, except in specified instances. LUBA frequently requests parties to waive the 77-day deadline and may enter an order on its own motion extending its decision period.

LUBA may affirm, reverse, or remand respondent's decision. Remand is more common than reversal, as the latter indicates the decision cannot be cured by further proceedings. The board will reverse or remand the land use decision under review if it finds that the local government, state agency, or special district exceeded its jurisdiction; failed to follow the procedures applicable to the matter before it in a manner that prejudiced the substantial rights of the petitioner; made a decision not supported by substantial evidence in the whole record; improperly construed the applicable law; or made an unconstitutional decision.

The most common grounds for review are improper construction (or misinterpretation) of the applicable law and making a decision not supported by substantial evidence in the whole record. Respondent is generally required to explain its decision in terms of what facts it believed, how it construed the law, and how it arrived at its decision in the light of the facts found and the law it construed. Moreover, where there are contested facts (e.g. whether a use is a "customary farm practice," or whether the level of service at an intersection is at a certain level), there must be evidence that a reasonable person could support the decision that was made after review of both supporting and opposing evidence.

Jurisdiction is rarely an issue before LUBA. Errors in notice or description are often viewed by LUBA as merely procedural, so that the petitioner bears the burden of demonstrating how the error prejudiced his or her substantial rights. Similarly, constitutional issues rarely occur at LUBA, though the board has the exclusive authority to review these in the context of a land use decision.

Judicial review of LUBA orders may be sought in the Oregon Court of Appeals. This review is confined to the record before LUBA and the court may reverse or remand LUBA's decision only if it finds the order: 1) unlawful in substance or procedure and prejudicing substantial rights of the petitioner; 2) unconstitutional; or 3) not supported by substantial evidence in the whole record as to facts found by the board. If the Court of Appeals reverses or remands LUBA's order, LUBA must respond to the court's opinion within thirty days.

Local Decision Making

Procedural protections in local land use decision making were first recognized by the Oregon Supreme Court in *Fasano v. Washington Co. Comm.*[10] In *Fasano*, the court ruled that the rezoning of a limited geographic area was a quasi-judicial act rather than a legislative act.[11] In *dicta*, the court noted that participants in a quasi-judicial proceeding are due certain procedural rights, including "an opportunity to be heard, an opportunity to present and rebut evidence, to a tribunal which is impartial in the matter . . . and to a record made and adequate findings executed."[12]

Since the *Fasano* case, notice and hearing requirements have been embraced, expanded, and spelled out in greater detail by the courts and the legislative assembly. Today, the governing body of each locality must adopt procedures for the conduct of hearings on quasi-judicial decisions consistent with those requirements.[13] Only ministerial decisions may be exempted from notice and opportunity for hearing requirements.

The issue of what constitutes a quasi-judicial decision has remained controversial since the *Fasano* decision. However, the focus of the controversy has shifted from the distinction between legislative and quasi-judicial acts to that between quasi-judicial and ministerial acts. This is due to the fact that, since the adoption of statewide goals by LCDC, there is little legislative land use decision making at the local level, as the adoption or amendment of most local land use plans and ordinances require the application of LCDC goals, thus constituting quasi-judicial decision making.

Ministerial decisions are those which involve no discretion on the part of the decision maker. These decisions simply require the application of clear and objective standards. An example is the determination that lots created by a subdivision comply with the required minimum lot area. In 1986, in *Doughton v. Douglas County*,[14] the Oregon Court of Appeals applied a narrow interpretation of what constitutes a ministerial decision and in the process brought a great deal of local land use decision making within the ambit of statutory notice and hearing requirements. In *Doughton*, the court found that the county's determination that a proposed dwelling was a "dwelling . . . customarily provided in conjunction with a farm use," constituted the exercise of discretion because it depended on "facts ... which are not determinable by simple reference to general provisions of the [county's] ordinance."[15] Therefore, the court ruled that the issuance of a building permit for the proposed dwelling was not a ministerial decision.

The *Doughton* opinion and several subsequent LUBA decisions have had significant implications for local decision makers.[16] Many determinations once thought of as ministerial are being treated as discretionary and therefore afforded all the procedural safeguards required by statute. While ensuring greater opportunity for affected persons to have input into the decision-making process, this broad interpretation has slowed down and increased the cost of the local permit process.

The permit process

Each county and city is required to adopt procedures for quasi-judicial decision making. State statutes provide general guidelines and allow localities to tailor their procedures to meet local needs.

The locality must provide a consolidated procedure by which an applicant may apply at one time for all permits or zoned changes needed for a development project. Typically, the application will be reviewed by planning staff and a formal report and recommendation will be submitted to the decision maker.

With some exceptions, the application must receive at least one public hearing. The procedures governing the conduct of quasi-judicial hearings are set forth by statute. Notice of the hearing must be provided to the applicant and to the owners of property located within a specified distance of the affected property.

Testimony and evidence at the hearing must be directed at the appropriate decision criteria. All parties have the right to rebut the evidence and testimony presented. Failure to raise an issue with sufficient specificity to afford the decision maker and the parties an opportunity to respond precludes an appeal to LUBA based upon that issue.

Approval or denial of an application must be based on standards and criteria set forth in the comprehensive plan, zoning and other implementation ordinances, and other local regulations, and be supported by findings based on evidence in the record of the hearing.[17] Regarding the adequacy of findings in quasi-judicial land use hearings, the Oregon Supreme Court has noted:

> No particular form is required, and no magic words need be employed. What is needed for adequate judicial review is a clear statement of what, specifically, the decision-making body believes, after hearing and considering all the evidence, to be the relevant and important facts upon which its decision is based. Conclusions are not sufficient.[18]

Subsequent opinions have emphasized that findings must "(1) identify applicable criteria, (2) find facts pertinent to those criteria, and (3) contain an explanation of the rational nexus between the facts, the criteria, and the result."[19] Written notice of the approval or denial must be given to the applicant and all parties to the proceeding.[20]

By statute, approval or denial of a land use application must be based on the standards and criteria that apply when the application is first submitted, if it is complete when submitted, or if the applicant submits additional requested information within 180 days after the application was filed. This provision is intended to prevent a locality from tightening or relaxing standards for approval of a pending application.[21]

In an effort to simplify the land use decision-making process for certain classes of proposals, the 1991 legislative assembly enacted legislation creating a new category called a "limited land use decision" which is not subject to the quasi-judicial procedures.[22] A limited land use decision is defined as:

> A final decision or determination made by a local government pertaining to a site within an urban growth boundary which concerns:
>
> (a) The approval or denial of a subdivision or partition . . . (b) The approval or denial of an application based on discretionary standards designed to regulate the physical characteristics of a use permitted outright, including but not limited to site review and design review.[23]

For limited land use decisions, the local government must provide notice of the proposal to owners of nearby property and any recognized neighborhood or community organization. Citizens then have fourteen days to submit written comments on the proposal. The approval or denial of the proposal must be based upon and accompanied by a statement of criteria and relevant findings and a notice of decision must be sent to the applicant and any person who submits comments.[24]

Discretionary approval or denial of a proposed development may be made without a hearing if notice of the decision is provided to all persons who would otherwise be entitled to notice of a hearing and an opportunity for appeal to the planning commission or governing body is provided. The appeal must be *de novo*, that is the matter must be addressed anew as if no prior decision on the proposal had been rendered.

In addition to providing for local review of decisions made without hearings, localities may establish procedures for the review of action of a hearing officer or other decision maker. The locality may prescribe that the planning commission or governing body or both are to hear such appeals. However, such internal review procedures are generally not mandated by statute and the governing body may provide that the determination of the original decision maker is final.

Regardless of what decision-making procedures are adopted by a locality, final action on an application for a permit or zone change, including the resolution of all local appeals, must be taken within 120 days after the application is completed. If the locality fails to take final action within the 120-day period, the applicant may file for a writ of *mandamus* to compel the governing body to issue the approval. The governing body can prevent the writ from being issued only by showing that the approval would violate a substantive provision of the local comprehensive plan or land use regulations.

Conclusion

The beginning of this chapter noted the happy coincidence of a number of political and legal currents that helped to give birth to the Oregon planning program. The consensus that each local government must have a binding comprehensive plan to provide the basis for further regulation is not a majority view in the United States, even today. To these views, Oregon added the notion of a state agency which would promulgate binding standards, against which city and county comprehensive plans and regulations would be weighed by that same agency. Enforcement authority was also provided to that agency to assure that state policy was not frustrated. As time went by, review of amendments to those plans and regulations and overall review of the entire package of the local government plan and its implementing regulations were added. Finally, an adjudicative body to weigh challenges to local government planning actions was added. Few states have more than two of these features; none but Oregon has all of them.

For planners commencing their careers in Oregon or coming to the state from elsewhere, the Oregon planning program often appears as a bewildering mass of agencies and regulations which require much study and experience. Assistance comes from their peers, from DLCD, and from their professional organizations, such as the Oregon Chapter of the American Planning Association. Some planners find the system too law oriented, rather than oriented to their own planning discipline. Nonetheless, most planners would agree that the relative complexity of the system is an adequate tradeoff for the certainty provided in the decision-making process.

Notes

1. 264 Or. 574, 507 P2d 23 (1973).
2. 271 Or. 500, 533 P2d 772 (1975).
3. *Id.* at 507.
4. The differences between compliance and compatibility are unexplored by caselaw.
5. ORS 197.015(5).
6. *Rajneesh Travel Corp v. Wasco County*, 13 Or. LUBA 202, 210 (1985).
7. ORS 197.190(3). To date, no such regional planning authority has been established under this provision.
8. *Id.*
9. The provisions of Ch. 804 Or. Laws 1993 (Enrolled SB 122) also provide for special district coordination with local governments and provide a remedy through an LCDC enforcement order for failure of a special district to do so.
10. Note 1, *supra*.
11. *Id.* at 581. (A quasi-judicial act involves the application of general policies contained in the comprehensive plan; a legislative act involves the formation of policy.)
12. *Id.* at 588.
13. ORS 215.412 and 227.170.
14. 82 Or. App. 444, 728 P2d 887 (1986).
15. *Id.* at 449.
16. See *McKay Creek Valley Assn. v. Washington County*, 18 Or. LUBA 71 (1989) (determination that proposed dwelling is "customarily required to conduct the proposed farm use" is a discretionary decision) and *Nicolai v. City of Portland*, 18 Or. LUBA 168 (1090) (city approval of minor land partition was a discretionary decision).
17. *Fasano v. Board of County Commissioners, supra*, note 1 at 588.
18. *Sunnyside Neighborhood Assn. v. Clackamas County Commissioners*, 280 Or. 3, 569 P2d 1063 (1977).
19. 2 *Land Use*, note 22, *infra*, sec. 35.35, citing *Lee v. City of Portland*, 57 Or. App. 798, 805, 646 P2d 662 (1982); *Dougherty v. Tillamook County*, 12 Or. LUBA 20, 31 (1984).
20. ORS 215.416(10) and 217.173(3).
21. Compare *Kirpal Light Satsang v. Douglas County*, 96 Or. App. 207, 772 P2d 1346 (1989) with *Sunburst II Homeowners Assn. v. City of West Linn*, 101 Or. App. 458, 790 P2d 1213 (1990). In *Kirpal*, the Court of Appeals held that

once an application is filed, the parties are bound to the standards and criteria in existence at the time of filing, rather than subsequently adopted standards and criteria. In *Sunburst II*, the Court of Appeals found no violation of the provision when, after LUBA found a city proposal inconsistent with City code provisions, the City amended the code, submitted a new application to itself, and approved the same. This, of course, was why the Court of Appeals affirmed the City's action of allowing a new application to be submitted and reviewed by a different standard. But see *Territorial Neighbors v. Lane County*, 16 Or. LUBA 641 (1988) and *Gilson v. City of Portland*, XXX Or. LUBA XXX (LUBA No. 91-93, November 15, 1991).

22. 1991 Or. Laws Ch. 814 14.

23. *Id.* at sec. 1.

24. ORS 215.416(11) and 227.175(10).

References

Sullivan, Edward J., and Larry Kressel. 1975. "Twenty Years After: Renewed Significance of the Comprehensive Plan Requirement." 9 *Urban Law Ann.* 33.

Sullivan, Edward J. 1990. "Growth Management and Land Use; The National Context." Unpublished manuscript.

CHAPTER 4
Irreconcilable Differences: Economic Development and Land Use Planning in Oregon

Matthew I. Slavin

The link between economic development and land use planning is close. A community's ability to capture a share of local and regional growth depends upon its ability to compete with other communities on the locational attributes and price of space. Land use planning plays a key role in determining this competitiveness.

Oregon has acknowledged the link between economic development and land use planning in its comprehensive statewide growth management system. The Oregon system is driven by nineteen goals, each of which serves as a general standard for local land use planning activities. Goal 9 mandates that local land use plans include provisions to promote development and diversification of Oregon's economy. In practice, however, Oregon has manifested this linkage between economic development and land use planning only perfunctorily. This chapter attributes the weakness of this link to regional disparities in the level of political support for economic development and land use planning, to differences in the state's economic development and land planning agencies' perceptions of their mandates to plan, and to the fragmentation characteristic of Oregon's system of state government.

The Politics of Oregon's Planning Mandates

Oregon's landmark system of statewide land use planning is a product of the 1973 session of the Oregon legislature, which created the Land Conservation and Development Commission (LCDC) and charged it with developing statewide growth management goals. One of the driving forces behind the act was the heavy development pressure bearing upon Oregon's Willamette Valley. Employment in the valley expanded by 43 percent between 1965 and 1973, compared to 28 percent in the nation as a whole. Urban sprawl threatened thousands of acres of valley farm and forest land, raising fears, as John DeGrove (1984) has put it, that a "tide of urban development would eventually wash over the Willamette Valley," turning a "natural paradise into a polluted nightmare."

Similar but less evident pressures also affected parts of coastal, eastern, and southern Oregon during this period. Whereas the Willamette Valley was home to a diversified economy composed of manufacturing and service industries, government, education, and agriculture, economic activity elsewhere in Oregon focused largely upon what Hibbard (1989) terms single staple industries—fishing along the coast, irrigated farming and ranching in eastern Oregon, logging all over but especially in southern Oregon. These areas trailed the Willamette Valley by almost every measure of growth and economic development. Consequently, residents of Oregon's less developed regions saw their public policy priority not as controlling growth but as closing the prosperity gap separating them from Willamette Valley residents.

In the Willamette Valley, the growth management program was viewed as a tool with which to "guide, direct, and control the quality of growth" (DeGrove 1984). Elsewhere, it was viewed differently. Throughout much of coastal, eastern, and southern Oregon, it was seen as an intrusive measure likely to inhibit prospects for local economic development and therefore was viewed with hostility. A breakdown of the legislature's vote on SB 100 helps make this point. Lawmakers representing the Willamette Valley voted for SB 100 by a 5-to-1 margin. In contrast, lawmakers from coastal, southern, and eastern Oregon opposed the bill 2-to-1 (Medler and Mushkatel 1979).

Senate Bill 100 was not the only mandate for planning to emerge from the 1973 session of the Oregon legislature. The session also produced Senate Bill 224, investing Oregon state government with responsibility for statewide economic development planning.

It was in areas hostile to Oregon's growth management act that support for SB 224 was strongest. The driving force behind the measure was State Senator John Burns. He was a Portland-area Democrat, but with strong ties to rural, conservative Oregon. Burns had been raised in eastern Oregon's rural Gilliam County, where both his father and brother held public office. Between 1969 and 1972, Burns served two terms as president of the Oregon Senate. Relations between Burns and his fellow Portland-area Democrats were often rocky, so much so that they attempted to depose him as Senate president in 1971. In a Senate chamber evenly divided between Democrats and Republicans, he was elected president only with the support of largely rural conservative Republicans. Burns's affinity with rural conservatives does much to explain the succession of votes he cast against SB 100 during the 1973 legislative session. He shared their view of growth management planning as an intrusive impediment to economic growth.

In June 1972, Burns, acting as Senate president, appointed a Senate Task Force on Economic Development. He was joined on the task force by four other state senators representing lesser developed areas of Oregon, including senators Ken Jernstedt of Hood River and George Wingard of Eugene, both conservative Republicans who cast key votes against SB 100. This trio sponsored SB 224, the economic development act.

A key aim was to reduce the influence of the governor's office in the sphere of state economic development planning.[1] The bill created the Oregon Economic Development Commission (OEDC), whose five members were nominated by the governor but required confirmation by the Senate. The bill also created the Oregon Economic Development Department (EDD), whose director was to be appointed not by the governor but by the commission. Senate Bill 224 directed the OEDC to establish a "comprehensive plan for the balanced community and economic development of the state." The plan was to be implemented by Oregon's new Economic Development Department. SB 224 clearly bore the stamp of lawmakers from Oregon's less developed regions. The bill mandated that in preparing a statewide economic development plan, EDD give "particular recognition to the needs, problems and resources of the rural and underdeveloped areas of the state."

Senator Burns was a strong opponent of the Oregon Land Use Planning Act. This highlights a certain irony to the SB 224 legislation: that Burns, an avowed opponent of state planning interventions, sponsored a bill that established a mandate for planning Oregon's economy. In fact, this result was not foremost on the Senator's mind when he established a senate task force. He announced his aim as the creation of an industrial development authority which could provide low-cost business financing in Oregon's less developed areas. The mandate for economic development planning in SB 224 can perhaps be best seen as the product of an accommodation between the legislature's rural growth contingent and growth management proponents.

Hearings conducted by Senator Burns's task force cast significant doubt upon the likelihood that a program of low-cost business financing could alone remedy the problem posed by underdevelopment in Oregon. Instead, it became apparent that much of coastal, eastern, and central Oregon lacked the physical infrastructure and the administrative, technical, and educational capacity to support the level of business activity necessary for economic diversification. The task force concluded that a reconfigured state role in economic development needed to provide a mechanism to channel state infrastructure assistance to needy communities.

Oregon's growth management proponents also realized that development required adequate infrastructure. Many of them viewed the expanded state role in economic development initially envisaged by Senator Burns as containing few assurances that resulting development would not be haphazard and hence incongruous with the intent and practice of SB 100. Growth management sentiments were especially strong on the Oregon House of Representatives' Committee on State and Federal Affairs, to which SB 224 was referred under the chairmanships of Portland-area Democrats Les AuCoin and Earl Blumenauer.

Still, there were limits on the extent to which lawmakers favoring growth management could oppose expansion of the state's role in economic development. Oregon's business and building trades, whose support was critical for SB 100, strongly supported an activist state economic development role. Furthermore, the period leading up to 1973 was one in which unemployment rates in the Willamette Valley persistently exceeded the national average. This was not a sign of fundamental economic weakness, for this was a period of rapid growth for the region. Rather, the coincidence of growth and relatively high unemployment was a sign that job creation simply could not keep pace with a more rapidly expanding population. High unemployment rates and strong business sector support made it difficult for proponents of growth management to oppose an expanded state role in economic development. Yet they did not want to license the state to engage in a program of unfettered growth promotion.

It was these circumstances that led SB 224 to cast EDD's role in terms of statewide economic development planning. So constructed, the bill satisfied Senator Burns and his allies by providing a mandate for the state to install the infrastructure necessary to promote rural growth. For their part, proponents of growth management saw SB 224 as an assurance that state-sponsored development would not be haphazard and therefore found themselves able to support the legislation.

Dual Planning Cultures

Opponents of statewide growth management successfully thwarted Governor McCall's efforts to secure funding for statewide land use planning during the 1973 session of the Oregon legislature. However, a special legislative session in 1974 afforded him another opportunity to seek these funds, this time successfully. The process by which LCDC proceeded to establish Oregon's statewide planning goals has not been extensively studied. Still, it is clear that the goals embodied several fundamental tenets.

Above all, LCDC's planning mandate was cast in terms of land conservation. This is most apparent in the case of the two most important and controversial of Oregon's statewide planning goals. These pertain to agriculture (Goal 3) and urbanization (Goal 14). The former, aimed at protecting Oregon's farm and forest resources, mandated the creation of exclusive farm and forest zones. The latter required the creation of urban growth boundaries, outside of which rigorous limitations upon property development were to be enforced. The purpose was to limit sprawl. Goal 9, relating to Oregon's economy and of key interest here, also serves to illustrate this point. Goal 9 requires that local comprehensive plans include an economic development element, the principal purpose of which is to set aside a supply of developable commercial and industrial land adequate for meeting anticipated future needs.

Oregon's statewide planning goals also evinced a regulatory role for LCDC, reflected in its responsibility for acknowledging that local plans were in compliance with the provisions of SB 100. To get their plans acknowledged, local jurisdictions have had to demonstrate that their planning decisions are based upon systematic and rational calculations. The LCDC recommended that local governments adhere to certain guidelines in seeking to justify their planning decisions. As regards Goal 9, LCDC's guidelines include "the health of the current economic base, materials and energy availability, labor market factors, current market forces, land availability, and pollution control requirements." The guidelines promulgated by LCDC are, in fact, voluntary. Still, local jurisdictions have largely adhered to them and the guidelines have afforded LCDC a yardstick by which local compliance with Oregon's planning goals can be measured.

It is clear as well that LCDC intended for its planning goals to be considered not individually but comprehensively. Goal 9 provided that decisions on setting aside land for future business development be made in conjunction with other planning goals addressing the provision of public infrastructure (Goal 11) and transportation (Goal 12). The aim was to ensure that land set aside for business development would be adequately served with infrastructure and transportation facilities.

The period leading up to 1973 often saw urban sprawl disjoin the type of land holdings best suited for large-scale industrial and commercial development in Oregon. In consequence, large-scale business development was often relegated to sites lacking infrastructure and services. This had emerged as a key concern of the Association of Oregon Industries (AOI), Oregon's main business lobby, which played a key role in shaping SB 100. The association envisaged the provisions of Goal 9 as a remedy to this problem, an instrument for ensuring that developers of business properties would have access to optimally sited and serviced land. This view was shared widely enough that the word "Development" was included in the name of the commission created to guide the implementation of SB 100. Whatever the degree to which LCDC's planning goals embodied a development ethic, however, was subsumed within the dominant context of land conservation. Indeed, the defining feature of the Goal 9 planning mandate was, above all, the need to reserve land for future development.

If the period 1973-76 saw progress in transforming SB 100's growth management planning mandate into practice, the same could not be said for Oregon's other mandate for statewide planning. Lawmakers from Oregon's less-developed regions remained strong proponents of an aggressive state posture on economic development. However, their enthusiasm was unmatched by proponents of growth management who, by 1975, had emerged ascendant in the legislature. Nor did Governor McCall demonstrate much enthusiasm for an activist state role in economic development. He declined to seek funding for EDD's planning effort during the 1973 and 1975 sessions of the Oregon legislature. For its part, the Economic Development Commission, nominated by Governor McCall and confirmed by a pro-growth management legislature, did little to indicate that it viewed the EDD planning effort as a priority.

Two events transformed the prospects of the economic development plan. One was the oil shock of 1973, which jolted the country into recession. As construction activity ground to a halt, the demand for Oregon's wood products plunged. The state's unemployment rate soared, peaking at 11.2 percent in July 1975, almost three points higher than the national average. The downturn fueled calls for a more activist state role in economic development, not only in Oregon's less developed regions but in the Willamette Valley as well.

A change in governors also affected the economic development plan's prospects. A constitutional provision limits Oregon's governors to two terms in office. Governor McCall's second term expired in January 1976. He was succeeded by former state treasurer Bob Straub. Governor Straub supported Oregon's growth management program. But his support was less ardent than McCall's, a point brought home by the *Oregonian*, Oregon's largest newspaper. It characterized Straub as "more of a developer than a preserver" and someone who "often grumbled at bureaucratic road blocks thrown up by agencies of the environmental movement" (October 1, 1978). Assuming office as Oregon's governor, Straub articulated his developmental proclivity, promising to "lure industry to Oregon" (*Willamette Week*, January 10, 1977).

The upshot of these circumstances became apparent in mid-1976. A resignation created a vacancy in the top EDD spot. For the new EDD director, Governor Straub sought someone he believed would act aggressively in expanding the state's economic development role. However, under the terms of SB 224, it was the Economic Development Commission, not the governor that was empowered to appoint EDD's director. Dominated by McCall nominees, the commission balked at Straub's choice. Straub responded by replacing recalcitrant commissioners. This enabled him to install his own choice as EDD head. Subsequently, Straub directed EDD to proceed with preparation of the statewide economic development plan authorized under SB 224.

The sponsors of SB 224 viewed government regulation as an impediment to economic development and the bill specifically proscribed EDD from any regulatory role. Reflecting the pro-growth sentiments of the bill's sponsors, EDD's role was cast in explicitly promotional terms. Furthermore, SB 224 eschewed any formal planning guidelines. In effect, the bill's planning mandate was open ended. This is in stark contrast to the growth management approach of SB 100, which established a rather specific set of rules which were to guide relations between LCDC, driven by a regulatory ethic revolving around land conservation, and local governments.

The different planning approaches embraced by LCDC and EDD during the 1970s were in part due to the influences of two individuals who played critically important roles during the early years of the two organizations. At LCDC, this individual was L.B. Day. Appointed LCDC's first chairman in 1973, Day presided during the period when Oregon's statewide planning goals were promulgated. At EDD, this person was Dan Goldy. It was Goldy whom Governor Straub fought to install as EDD director in June 1976.

Prior to becoming LCDC head, Day had served as director of Oregon's Department of Environmental Quality. He was also an official of Oregon's Teamsters Union, a position he continued to hold while chairing the commission. As a lobbyist, he played a key role in shaping SB 100 during the Oregon legislature's 1973 session. At LCDC, Day strongly supported the land planning mandate he had helped craft. Day was a brusque individual, prone to caustic speech and reluctant to compromise, qualities which almost certainly hastened his departure as LCDC chair in 1976. The organization he helped to build, though, had already earned a reputation for zealous enforcement of Oregon's statewide planning protocols.

Dan Goldy was a former deputy assistant commerce secretary in the Kennedy administration, former vice-president of the U.S. Chamber of Commerce, and past president of a large Texas corporation. While Day was often gruff and uncompromising in defending his vision of the Oregon land planning program, Goldy was politically adroit and polished. He infused EDD with an entrepreneurial spirit that led University of Oregon economics professor Ed Whitelaw to characterize the agency as a "statewide Chamber of Commerce" (*Willamette Week* May 1, 1978). Commenting on Goldy's entrepreneurial abilities, Oliver Larson, executive vice-president of the Portland Chamber of Commerce, said: "He could brush his teeth and bring someone into the state" (*Willamette Week* January 10, 1977). Goldy's involvement with EDD began auspiciously enough. One of his first acts as EDD director was to secure a $1.5 million grant from the U.S. Economic Development Administration for use in preparing the statewide economic development plan mandated under SB 224.

Irreconcilable Differences

The 1973 session of the Oregon legislature had produced two planning initiatives that addressed economic development. Senate Bill 224 mandated creation of a statewide economic development plan. Senate Bill 100 required the inclusion of economic development elements into local comprehensive planning documents. However, the regionally divided legislature produced no edict for harmonizing these planning mandates.

Despite the lack of legislative direction, an opportunity to harmonize the planning mandates of SB 224 and SB 100 nonetheless arose in the form of a decision by EDD to adopt a two-track approach to its responsibilities. The department decided to ask other state agencies such as the agriculture, energy and transportation departments to draft documents outlining the roles they could play in promoting development of Oregon's economy. EDD intended to incorporate these documents, as well as the economic development plans of Oregon's city and county governments, into a master statewide economic development plan. The department thereby aimed to establish a concerted regime for mobilizing public and private resources in pursuit of economic growth.

At the time, economic development planning was new to many of Oregon's county and city governments. In many cases, economic development planning simply meant complying with LCDC's Goal 9 provisions. As a practical matter, EDD's decision raised the prospect of coordinating the Goal 9 mandate with the SB 224 mandate. This prospect did not go unnoticed by EDD, which inaugurated a local assistance program. John Mosser, L.B. Day's replacement as LCDC Chairman, agreed to review EDD's local assistance program for approval for use by local governments in Goal 9 planning efforts.

However, these initial efforts at harmonized planning were plagued by differences in the level of local government at which EDD and LCDC aimed their planning interventions. While EDD offered assistance to all local governments, it preferred to work with regional economic development districts. This preference was not arbitrary. The districts were eligible for funds from the U.S. Commerce Department's Economic Development Administration, the same funding source that enabled EDD to undertake its statewide planning effort in the first place, and possessed experience in the economic development field that many municipal and county governments lacked. The districts served mainly as vehicles for promoting business development; they shared with EDD an entrepreneurial orientation focused upon growth promotion. EDD's preference for working with economic development districts conflicted with DLCD planning, since it was not the districts but Oregon's city and county governments that were responsible for complying with the Goal 9 provisions.

Efforts to coordinate the SB 224 and Goal 9 mandates also suffered from disputes over the proper basis for planning decisions. The guidelines LCDC established for use by local jurisdictions in complying with Goal 9 did not always square with EDD's planning methodology. For example, LCDC guidelines required local jurisdictions to base their Goal 9 planning decisions upon twenty-year projections of employment growth. EDD utilized a different set of analytical criteria based upon its own employment growth projections and growth rates in different industrial sectors. Perhaps the most visible evidence of the conflicts over planning assumptions was LCDC's ultimate refusal to certify EDD's local assistance program for use in Goal 9 planning.

A third problem in achieving coordination involved basic differences in values between EDD and LCDC. Writing in 1984, John DeGrove stated that "Goal 9 is an example of a goal that has been neglected by the LCDC" (DeGrove 1984, p. 275). While Goal 9 ascribed a role in economic development planning to LCDC, the commission did little to demonstrate that it viewed this as a priority. The commission's attention focused mainly upon ensuring local government compliance with goals more central to its perceived mission of land conservation, especially the establishment of urban growth boundaries and farm and forest land preserve zones. One consequence was conflict between EDD and LCDC over how much of a priority local planners should attach to the economic development components of their comprehensive plans.

Many of the planners responsible for bringing local comprehensive plans into compliance with Oregon's statewide growth management planning goals were employed under the aegis of LCDC grants (DeGrove 1984). Consequently, they tended to attach greater legitimacy to LCDC's directives than to those promulgated by EDD, which encountered difficulties in getting local planners to engage in the type of planning upon which it was counting for completion of the statewide economic development plan.[2] EDD's options for engendering cooperation by local authorities were limited. The promotional approach established by SB 224 provided EDD no enforcement authority akin to that wielded by LCDC.

Finally, economic recovery undermined the effort at harmonizing the SB 224 and Goal 9 mandates. Oregon's unemployment rate dropped rapidly after peaking in 1975. By 1977, Oregon's economy was again growing more rapidly than that of the nation as a whole. Recovery eroded public support for an activist state role in economic development. Concomitantly, support for statewide growth management planning was reinforced, as demonstrated in referendums held in 1978 and 1980. Still, these ballots showed that support remained strongest in the Willamette Valley area. Elsewhere, growth management planning continued to encounter hostility, in some cases overcome only by direct LCDC intervention and enforcement actions.

The Demise of Harmonized Planning

Taken together, these circumstances undermined efforts to coordinate Oregon's SB 224 and Goal 9 planning mandates. For growth management planning in Oregon, this failure had marginal consequences. Goal 9 was, after all, but one component of a comprehensive planning regime and, in LCDC's view, not a high priority. With strong public support and an enforceable mandate, statewide growth management planning became institutionalized as a fundamental tenet of public policy. Bringing all local planning jurisdictions into compliance with SB 100 proved slow going: LCDC initially set January 1, 1976 as the deadline and a series of extensions put this deadline off for four years. Still, by 1980, all but 56 of 266 local planning jurisdictions had submitted plans for acknowledgement (the *Oregonian* September 15, 1987).

A successful joint effort was more critical to EDD since the agency envisaged local economic planning elements as a major component of its statewide master plan. No economic development plan was ready for the 1979 session of the legislature, despite a series of promises. Frustrated with a master planning process that may have consumed in excess of $1 million but produced meager results, lawmakers voted 67 to 10 to repeal the provision of SB 224 that had originally directed EDD to engage in statewide economic development planning, killing the project.[3]

Could Oregon have overcome the problems that impeded harmonization of the SB 224 and Goal 9 mandates? Certainly some problems transcended administrative solution. For example, harmonization succumbed to tensions founded in levels of public support that varied over time and place with economic conditions largely dictated by forces beyond Oregon's control. Other problems were potentially more tractable. For example, in some cases, the economic development districts that EDD preferred to work with existed as parts of regional councils of government (COGs). As originally drafted, SB 100 would have made Oregon's COGs, not cities and counties, responsible for local implementation of the state's growth management planning system. Adoption of this original scheme would have helped unify the implementation mechanism of these dual planning mandates. However, the COG scheme for implementing SB 100 was abandoned in the face of opposition from county and municipal government leaders (Leonard 1983). They saw COGs as "nonlocal intervention by nonelected officials" and feared that giving COGs SB 100 planning responsibilities would usurp authority they themselves enjoyed. As a practical matter, their parochial politics foreclosed a prospective avenue for facilitating harmonization of the SB 224 and Goal 9 mandates.

There was also the problem of fragmentation. As a consequence of longstanding historical factors, authority in Oregon state government is highly fragmented, with public authority divided between the governor, legislature, and numerous quasi-autonomous administrative and policy-making bodies. This arrangement maintains a balance of power between different components of Oregon's polity. However, it tends to weaken the authority necessary for resolving conflicts arising between these bodies. Perhaps nothing illustrates this better than the case of the Economic Development Commission and LCDC. Authority for making and implementing state policy was, in each case, installed in a largely autonomous body only tangentially accountable to other elements of the state's policy-making apparatus.

Fragmentation has long been recognized as an obstacle to effective public policy action in Oregon. Again, Oregon's experience in seeking to harmonize the SB 224 and SB 100 mandates serves to illustrate this point. Take, for example Governor Straub's 1976 firing of economic development commissioners. This move was clearly made in an effort to centralize authority. Still, the prospects for extending Straub's success were small. The 1979 legislature attached to the bill repealing EDD's statewide economic development planning responsibilities a provision that effectively proscribed Oregon's governor from again removing state economic development commissioners from office for all but malfeasant offenses.

The Legacy of Two-Track Planning

If growth management dominated public policy debate in Oregon during the 1970s, economic development was the key public policy issue in Oregon during the 1980s. Mainly due to its continued dependence upon wood products, Oregon was hit unusually hard by the 1980-82 recession, registering the sixth-highest unemployment rate in the nation in December 1981, followed by an uneven recovery. Subsequent growth was concentrated in the urban communities of the Willamette Valley. Automation in the wood products sector, a farm crisis, and looming shortages in the supply of harvestable timber resulted in a severe prolongation of the downturn throughout much of coastal, eastern, and southern Oregon. In November 1985, twenty-five of Oregon's thirty-six counties, all largely timber dependent and located in coastal, southern, or eastern Oregon, were classified by the U.S. Labor Department as labor surplus areas, meaning that unemployment rates in these counties had exceeded the national average by 20 percent or more throughout the preceding 24-month period.

As in 1975, economic downturn fueled calls for a more activist state economic development posture. But this time the downturn was more severe and prolonged and the calls for action more vehement. Oregon's new Governor Vic Atiyeh (1979-86) made economic development his number one priority. The centerpiece of the Atiyeh administration's program was an aggressive campaign aimed at recruiting industry to Oregon. As regards statewide growth management planning, the Atiyeh administration adopted something of an equivocal posture. Governor Atiyeh had voted for SB 100 while representing west-side Portland in the 1973 session of the Oregon legislature. Senate Bill 100 came up for renewal in 1981, during his tenure as governor. Governor Atiyeh backed reauthorization of LCDC, siding against those who would have eviscerated or eliminated the body. He also sought a special appropriation from the Oregon legislature for activities aimed at inventorying lands available for commercial and industrial development. On the other hand, concomitant with its industrial recruitment activities, the Atiyeh administration mounted a campaign to improve Oregon's business climate. At the crux were moves to relax state tax and regulatory regimes. The LCDC emerged as a target of these activities. Reflecting sentiments not dissimilar to those long expressed by opponents of growth management planning, Governor Atiyeh criticized LCDC for impeding growth by being overly zealous in its regulatory responsibilities (the *Oregonian* May 20, 1982).

A majority of Oregonians viewed maintenance of a regulatory growth management planning regime as a necessary instrument for ensuring the state's livability, though some continued to view it as an impediment to economic development. As a practical matter, Oregonians were coming to view statewide growth management planning and economic development policy as potentially opposed areas of policy intervention.

The extent of the conflict became clear in the latter half of the 1980s. Continued divergence in the economic fortunes of the Willamette Valley and Oregon's less developed regions prompted a 1985 edition of the *Oregonian* to report "Oregon's economy is rapidly becoming two" (January 27, 1985). Not only did the Atiyeh administration's economic development program fail to stem this divergence; evidence suggested that it actually served to increase regional disparities.[4]

These circumstances created a crisis atmosphere, fueling calls for state economic development policy reform. From these emerged the Regional Strategies program, centerpiece of the economic development efforts of Governor Neil Goldschmidt, who succeeded Governor Atiyeh as Oregon's chief executive in 1987. Through Regional Strategies, Goldschmidt aimed to make good on his campaign promise to lead an "Oregon Comeback," which would extend recovery to "those counties and constituencies that are most in need of a revived economy" (the *Oregonian*, November 7, 1985). Reflecting the urgency attached to solving Oregon's development crisis, the 1987 session of the state legislature agreed to emergency authorization of the Regional Strategies program and appropriated $25 million in state lottery dollars for its funding.

Under the terms of the Regional Strategies program, Oregon's counties were directed to conduct analyses of their local economies. Each county was to use its analysis to formulate strategies for promoting local economic diversification. Subsequently, they were to submit to Governor Goldschmidt "wish lists," identifying actions they aimed to undertake in promoting diversification and, importantly, identifying forms of assistance they sought from various state agencies in order to implement their economic diversification strategies. The governor would then act to see that the requested assistance was provided.

The Regional Strategies program amounted to an attempt at statewide economic development planning—using analysis to establish goals and to direct the concerted application of available resources. It is clear, however, that Governor Goldschmidt envisaged Regional Strategies as a more structured and robust mandate than provided for under SB 224. The Goldschmidt administration sought from the 1987 session of the Oregon legislature a grant of extraordinary authority over several of the quasi-autonomous state commissions it viewed as having an important role to play in promoting economic diversification. This was intended to overcome the problem of fragmentation that had impeded earlier efforts.

The Economic Development Commission was one such commission. Reflecting the important role that he envisaged for roadway, water, and sewer infrastructure investments in his Regional Strategies effort, Governor Goldschmidt sought extraordinary authority over the state Transportation and Environmental Quality commissions. He also made it clear that he envisaged an important role for the Oregon State Marine Board, responsible for making port investments. In the event, the Oregon legislature declined to grant such extraordinary authority. It is instructive, however, to note that Governor Goldschmidt did not seek to extend his authority over LCDC. So pervasive was the separation between growth management and economic development planning that at a time when Oregon found itself resorting to extraordinary measures in confronting a development crisis, it did not even seek to enlist the state's growth management apparatus, with its Goal 9 economic development planning responsibilities. In retrospect, however, this should not be surprising. The Regional Strategies program was aimed, after all, primarily at promoting growth in Oregon's less developed regions, long the focal point of opposition to growth management planning.

Summary

Oregon's efforts to harmonize statewide economic development and land use planning activities have largely failed. This point is illustrated by Oregon's inability during the 1970s to connect its statewide comprehensive growth management planning activities with efforts to develop a statewide economic development plan. The effort was undermined by regional disparities in the level of political support for economic development and land use planning, differences in how Oregon's economic development and growth management planning agencies perceived their mandates to plan, and the fragmentation characteristic of Oregon state government. Consequently, Oregon's statewide growth management planning and economic development activities diverged. The extent of this separation became clear with the Regional Strategies program, the statewide economic development initiative launched in the latter half of the 1980s. So divorced had economic development and growth management planning become from each other that the architects of the Regional Strategies program did not even view Oregon's growth management planning apparatus, with its Goal 9 economic development planning mandate, as having an important role to play in redressing what was perhaps the state's greatest economic development challenge since the Great Depression.

Notes

This chapter is an outgrowth of research the author conducted while preparing a Ph.D. thesis in Urban Studies at the School of Urban and PublicAffairs, Portland State University (Slavin 1992).

1. Discussion here draws heavily upon records of the proceedings of the Oregon Legislature's Senate Task Force on Economic Development and of the proceedings of the 1973 session of the Oregon Legislature's Senate Committee on Economic Development and House Committee on State and Federal Affairs.
2. Interview with Dan Goldy, November 13, 1991.
3. An exact accounting of how much money was spent on EDD's local assistance/economic development planning effort is unavailable. The $1 million figure represents the author's estimate.
4. Industrial revenue bonds (IRBs) were the main instrument used by the Atiyeh administration to induce business investment in Oregon. Figures provided by the Oregon Economic Development Department indicate that Oregon issued a total of $267.43 million in IRBs between 1980 and 1985. Six Willamette Valley counties accounted for $167 million or 62.5 percent of this total: Multnomah, Washington, Clackamas, Marion, Benton, and Linn. Only $100.3 million, or 37.5 percent of IRB funding during this period, went to the counties located in eastern, southern, or coastal Oregon, where economic conditions were most distressed. Furthermore, of this $100.3 million, $4.7 million went in to Hanna Nickel in 1985, a Douglas County ore smelter which nonetheless closed down two years later. Another $31.5 million appears to have been used for mill retrofits which may have actually reduced employment in the mill sector.

References

DeGrove, John M. 1984. *Land Growth and Politics*. Chicago: American Planning Association Planners Press.

Hibbard, Michael. 1989. "Small Towns and Communities in the Other Oregon." In *Oregon Policy Choices: 1989*, edited by Lluana McCann. Eugene: Bureau of Government Research and Service, University of Oregon.

Leonard, H. Jeffrey. 1983. *Managing Oregon's Growth: The Politics of Development Planning*. Washington: The Conservation Foundation.

Medler, Jerry and Alvin Mushkatel. 1979. "Urban-Rural Class Conflict in Oregon Land Use Planning." *Western Political Quarterly* 32, 3: 338-349.

Slavin, Matthew I. 1992. *State Industrial Policy and the Autonomy of State Leaders: Evidence from the Oregon Experience*. Ph.D dissertation, Portland State University.

PART II
Applying the Oregon System: Planning Issues and Choices

CHAPTER 5
Housing as a State Planning Goal

Nohad A. Toulan

The state of Oregon's active involvement in statewide land use planning is only twenty years old. During this period the state has become a national model for innovative and forward-looking approaches to land use regulations and regional growth management policies. Oregon's positive image among planners is based more on its approach than on the simple fact that it was one of the first to recognize the need for statewide land use planning. In fact, the state is a relative newcomer when compared to states such as New Jersey and Wisconsin, where state planning efforts date back to the period between the two world wars. Oregon's contribution to planning, therefore, is based not on its length of involvement but rather on innovation. It established new directions for existing ideas, giving them reason and substance and creating appropriate mechanisms for enforcement and implementation. Some may fault the state for creating an elaborate set of land use regulations rather than a statewide development plan. However, those regulations were based on a comprehensive list of planning goals that included housing and other issues that are central to the economic and social well-being of all Oregonians. This aspect, when examined in the context of the 1970s, was an important evolutionary step along the continuum that marks the development of urban and regional planning in the United States. Housing in general and socially sensitive housing policies in particular became a distinguishing feature of the Oregon approach. There is no better way to dramatize the importance of this approach than to review the role of housing in the evolution of planning thought. Such a review will reveal that, while Oregon's land use planning process followed in the footsteps of an already emerging general trend, the state's goals for affordable housing and the way they were interpreted and enforced broke new ground.

Housing and the Evolution of Planning Thought

Of all the activities affected by land use planning and regulations, housing must be singled out as the most significant and far reaching in its impact on the form and structure of metropolitan regions. It is not only the largest consumer of land but is also the one activity which affects most if not all social institutions. In fact, it was concern with appalling housing conditions in the nineteenth-century industrial city that propelled the reform movements which laid the foundations for the rise of the planning profession in the United States. Awareness of the importance of housing was reflected in the ideals of early and middle nineteenth-century reformers such as Charles Fourier in France and Robert Owen and James S. Buckingham in England. Their writings leave no doubt about their recognition of the social significance of housing. Concern with urban housing conditions, however, predates the industrial revolution. A valid case can be made for the notion that housing as an important urban element is as ancient as cities themselves. Awareness of the negative impact of bad housing on the overall health and appearance of the city was clearly an issue in ancient societies (Mumford 1961, pp. 465-481). Responses to the challenge varied, but when a laissez-faire attitude prevailed it resulted in considerable peril to the health and welfare of the inhabitants.[1] History is replete with examples that leave no doubt about the damage to the urban fabric that results when housing conditions are ignored. Examples are extreme in the industrial cities of the nineteenth century and in the emerging metropolises of the Third World today.

In the United States, attempts by government to play an active role in improving urban conditions began by regulating housing (Scott 1969, pp. 1-10). While it was concern with public health and safety that motivated government to act, social workers and humanitarians who took the lead in exposing the deplorable conditions in the slums of our major cities were more concerned with tenants' welfare. Whether it was concern for public health or tenants' welfare that generated action is not as important as the fact that it was housing that brought government into the business of urban regulations.[2] The 1867 New York City Tenement Act is recognized as one of the first milestones in the evolution of the modern city planning regulatory process. The act was the first of a series that sought to curb exploitation and improve living conditions. These acts were much closer to building and design regulations than to contemporary housing legislation, but they reflected the type of social consciousness embodied in the writings of the late-nineteenth-century reformers. However, the nineteenth century, which brought us face to face with the socially degrading effects of slum housing, was still dominated by puritanical ideals that blamed the poor, at least partially, for their problems.

Reforms were directed towards improving living environments more than towards enhancing the availability of decent affordable housing. In other words, the symptoms of the housing problem and their adverse impacts were acknowledged but the solutions were often off the mark. These solutions favored aesthetic and design improvements and ignored the socioeconomic roots of the problem. As a result, while the problem was the availability of decent shelter, the solutions—as exemplified by the White City and the City Beautiful Movement that followed—were peripheral. But the shortcomings of the emphasis on aesthetics were not ignored. By the mid-1900s criticism of the City Beautiful Movement was gaining momentum on the ground that its schemes were superficial and obscured the larger intention of the reformers' efforts (Scott 1969, pp. 78-80; Klaus 1991, p. 460). Nevertheless, the emphasis on physical planning was to remain a major driving force shaping planning thought well into the second half of the twentieth century.

Until the early 1960s the tendency to separate the physical and social aspects of housing dominated our approach to the management of this important urban element. As a result, housing became secondary to other concerns, regardless of its centrality to all that makes a city liveable. It was addressed in the context of what it means to the urban pattern, economic development, transportation networks, or urban aesthetics but rarely as a separate element with independently significant social and economic merits. Nowhere was this situation more evident than in the concept of the master or general plan as introduced by Bassett and Bettman in the 1920s and strongly advocated by Jack Kent in the 1950s and 1960s (Bassett 1938, Bettman 1929, Kent 1964). Bettman and Kent, whose ideas dominated city planning thought until the end of the 1950s, were concerned with the relationship between the various elements of land use.[3] In this context the general plan was to deal with different land use categories, not with activities and certainly not with questions of social justice and equity. In other words, the housing element of the general plan was reduced to a blueprint for the location, size, and type of residential areas. In fact the term "housing" does not appear anywhere in Bettman's 1928 Standard Planning Enabling Act and was absent from most plans until the early 1960s.

Stuart Chapin presented a similar but slightly more balanced view of the nature of city planning in 1957. In answering the question, "What does contemporary city planning encompass?" he suggested that planning "may be regarded as a means for systematically anticipating and achieving adjustment in the physical environment . . . consistent with social and economic trends and sound principles of urban design" (Chapin 1957, p. xiv). He went on, however, to draw a clear distinction between a large, comprehensive city planning process and a more confined land use planning element. That distinction, while elementary by our current standards and understanding, was very significant under the conditions that prevailed at that time. He was also more balanced in his approach to the study of residential space requirements and introduced several assumptions that are, today, central to our ability to study housing.[4] In retrospect, this was a modest evolutionary milestone. However, Chapin's planning methodology, which was widely utilized at the time, was based on a definition of planning that was not much different from that embodied in the 1928 Standard Enabling Act.

Indeed, it could be argued that, throughout its earlier history, city planning did not concern itself with activities as much as with land use categories, and the omission of housing was no exception. Other activities such as industry and commerce were also treated as land use categories, with the emphasis placed on spatial relationships rather than on employment opportunities, type of jobs, and other socioeconomic concerns that we now believe should be considered when dealing with the siting of commercial and industrial uses. Nevertheless, the practical application of such an approach to housing had greater negative implications because it allowed planners to avoid addressing socially sensitive and politically controversial questions. It also meant that the one activity which consumes more urban land than any other and affects all residents was to be addressed in a superficial way without full understanding of the effects of alternative land use scenarios on our social and economic well-being.

The city-federal partnership

The fact that early attempts to institutionalize the planning process and the movement towards the standardization of its scope failed to acknowledge the social significance of housing does not mean that such issues and concerns were universally ignored. As already indicated, housing conditions in the slums of major industrial cities were the subject of intensive social debate throughout the second half of the nineteenth century and the early part of this century. The centrality of housing was also reflected in the writings of theoreticians such as Ebenezer Howard and Patrick Geddes who shaped planning thought in the early years of the profession (Howard 1902, Geddes 1915, Ch.VI). However, as planners, including those who accepted the importance of housing in shaping urban form, concentrated on the physical side of the planning process, concern with the socioeconomic aspects of housing began shifting to the federal government. As cities gained weight in the national political arena, improvement of low- and middle-income housing conditions and availability became a major factor influencing the evolution of federal urban policy and programs starting with those of New Deal in the 1930s. However, as long as urban planning was limited to policies and scenarios that focused on the future use of land, questions of equity and the proper functioning of the various urban subsystems remained of lesser concern to mainstream planners and were left for others to address. In other words, planners gave up their responsibility for a truly comprehensive planning process that addresses all aspects of urban form. In retrospect, that situation reflected the political realities of the time, with the federal government moving ahead of the states in espousing liberal causes.

Starting with the 1949 Housing Act,[5] Congress and a major segment of the federal establishment began advocating goals and values that gradually placed them in the vanguard of a social liberalization movement that peaked with the Great Society programs of the Johnson administration in the mid 1960s. It was a period during which the federal government, not state and local governments, set the agenda for urban reform, especially in the larger central cities where mayors developed a sense of close partnership with the federal government.

While the partnership between the federal government and the cities encompassed a wide array of programs, housing emerged as a central concern from the beginning of the relationship in the 1930s (So et al. 1986, p. 403). Housing legislation became the central piece in all urban action programs and, with the passage of every new housing act since the first in 1934, federal influence in shaping housing and urban development policies grew in scope and significance. Federal involvement led to a mistaken perception that ensuring equity in housing markets is a federal not a local responsibility.⁶ With Richard Nixon's election in 1968, the role of the federal government vis-a-vis local problems went through a slow process of change that gained momentum during the Reagan years. Cities could no longer count on the federal government to bail them out and most turned to local resources and initiatives. The cities were not totally unprepared. The 1973 Housing and Community Development Act, which was intended to simplify forty years of accumulated federal programs, had required cities and metropolitan areas to assume greater responsibility for planning and coordination. It mandated the development of housing plans and enhanced the role of citizen participation in the community development process.

Local change, however, was not entirely due to dictates from Washington. Two reform movements that began in the late 1950s and early 1960s converged in the 1970s to produce a new political environment conducive to greater involvement by the states. These were, on one hand, growing concern for the environment and, on the other, the social activism movement with its emphasis on the need to meaningfully engage the public in the decision-making process and the importance of addressing emerging social concerns. The first generated pressures on the states to become more actively engaged in land use issues, while the second meant that socially sensitive elements of the urban development process were given serious consideration. Housing figured prominently among those elements. The states regained the momentum they had lost to the federal government and, even though they lacked the resources, the mere fact that they were reengaged was a positive development.

Reengaging the states

During the 1970s environmental concerns focused on a wide array of issues ranging from air quality to species protection with the federal government still active as a key player. The states, though, were also on the move and through legislation or referenda began to engage in activities previously relegated to local or national agencies. The "Bottle Bill" pioneered by Oregon in the mid-seventies is a good example of state efforts to protect the natural environment. However, it was in the area of statewide planning and growth control that the changing role of the states was more profound. Again Oregon attracted the national attention through the "no growth" campaign of then Governor Tom McCall. That campaign, while misguided in regard to the nature of the problem, generated considerable interest in statewide land use planning and the question of growth management.[7]

Concern for the environment was not the only driving force behind the changes that occurred in the way we approached urban issues. Starting in the early 1960s the planning profession had begun to confront the need for a better understanding of social and economic issues with an eye towards equity and justice. More important perhaps was the emerging concern for an efficient and open process for planning implementation. In 1962, Paul Davidoff and Thomas Reiner brought to the forefront the question of values and their importance in the planning process. In "A Choice Theory of Planning" they argued that "the planner, as an agent of his [sic] clients, has the task of assisting them in understanding the range of the possible in the future and of revealing open choices. He does this in two ways—one involving facts and the other, values. . . . values are inescapable elements of any rational decision making process or of any exercise of choice. Since choice permeates the whole planning sequence, a clear notion of ends pursued lies at the heart of the planner's task" (Davidoff and Reiner 1962, pp. 107, 111). While their intent was to develop a general theory of planning that is universally applicable, they actually unleashed, along with John Dyckman, Herbert Gans and other social planners, a major shift in planning emphasis towards greater advocacy and higher levels of citizen involvement.[8] This shift and the impetus it gave to activism in the planning arena was a significant catalyst for emerging social and environmental concerns.

The movement to protect the environment was easier to incorporate in the mainstream of prevailing planning philosophies and was better able to build on past experiences with regional planning. It also benefited from close affinity to the open-space preservation ideas first presented by William Whyte in the *Exploding Metropolis* (Whyte 1958) and dramatized by Ian McHarg in *Design with Nature* (McHarg 1969). By the early 1960s some states were already developing programs to address the need for alternatives to uncontrolled urban expansion.[9] These programs, however, failed to fully address the social implications, particularly the impact on moderate- and low-cost housing. This does not mean that concern for housing was lacking. Quite to the contrary, as could be inferred from reviewing the literature of that period, the interest in fair housing policies was as great as or even greater than the interest in environmentally sensitive growth policies. What was missing was the link between the two. While by current norms the failure to tie the two issues could be inexcusable, it should be remembered that those were the days when the link between suburban expansion and low-income housing shortages in the central cities was difficult to make or substantiate; and when that link was made it was by social activists intent on liberalizing zoning regulations in the suburbs to provide greater opportunities for low-income families. Davidoff's Suburban Institute in suburban Bergen County, N.J., was an example of an organized effort to accomplish such goals.

The strong bilateral ties between the cities and the federal government tended to project an image of disengaged state governments, unconcerned with the problems afflicted on and created by the growth of their larger cities. Indeed, that was usually the case in situations where a state government was more conservative than its local and national counterparts. Obviously, there were exceptions and many of the states were gradually developing programs ranging from direct housing finance to enforcing fair housing regulations (So et al. 1986, p. 408), but it was the failure to establish the link between urban growth issues and housing that contributed to the perception of disengagement. In situations where housing policies were addressed within the context of the larger urban development perspective those perceptions were less extreme.[10] In reality, however, state planning activities, that started in earnest in the 1920s, declined in the 1940s, and regained slow momentum in the 1950s and 1960s did not deal directly with broad housing policy issues.

The situation began changing in the mid-1970s when urban growth management became a matter of emerging concern for the states. Coincidentally, that was also a time of considerable challenge for the cities. The federal government was on the retreat and local social activism, especially with regard to housing, was on the rise. As a result, it is difficult to determine the real motives that led the states to develop the connection between land use regulations and housing policies. Was it the product of a genuine interest in a comprehensive approach to growth management or an opportunistic response to the prevailing political realities? Debating this issue may be of some intellectual value but whatever conclusion is reached the fact remains that the outcome was significant. It was in that changing environment that Oregon began addressing the challenge of urban growth and the results reflected some sensitivity towards social implications. However, Oregon's move into the arena of statewide urban growth management caused some apprehension about possible side effects with negative impacts on the land and housing markets. As a result, Oregon's story is that of discovery in unchartered waters; high risk but worthy rewards.

Housing and Growth Management in Oregon

Oregon's involvement in regulating land use gained legislative support with the passage of Senate Bill 10 in 1969. The bill was referred to the voters by petition and affirmed in 1970. As a result all cities and counties were required to adopt and apply comprehensive planning and zoning ordinances. To assist in the development of such plans the bill established nine statewide goals. The bill's emphasis, as illustrated by those goals, was clearly on transportation and the physical environment. In fact, of the nine goals only one dealt with issues not directly related to land use. Goal 8 called for comprehensive physical planning to address the need "to diversify and improve the economy of the state" (SB 10, 1969). Housing and other socially significant elements of the urban environment were not addressed directly by any of the nine goals but were mentioned in the introduction. In fact, that introduction was one of the earliest definitions of comprehensive development planning adopted by a state legislature. It describes planning as a tool needed "to assist in attainment of the optimum living environment for the state's citizenry and assure *sound housing*, employment opportunities, educational fulfillment and sound health facilities" [emphasis added]. Viewed in retrospect the bill was, indeed, innovative and regardless of its many omissions it is often viewed as an important milestone on the road to the development of a comprehensive state housing policy[11] (Sullivan 1990).

The major responsibility for monitoring compliance was entrusted to the governor but without a clearly defined institutional framework other than the State Land Board which lacked a mandate for bold initiatives. The quickly recognized limitations of the bill led to adoption of Senate Bill 100 in 1973. Of the nineteen goals developed by the new Land Conservation and Development Commission (LCDC) following passage of the new bill, several are important to housing but only two have direct bearing on the operations of this sector.[12] Goal 10 deals with housing *needs* and was the first attempt by the State to inject the social side of housing concerns into the Oregon comprehensive planning process. Goal 14 has equal significance to housing since it addresses urban growth boundaries and the "realistic" availability of land for residential development. However, because it is primarily concerned with growth management it was viewed as impacting housing in the urban fringe. The impact, though, of urban growth boundaries on residential land values became a subject of intensive debate (Nelson 1984) and was the one aspect of Oregon's program that generated concern among housing advocates. This subject is examined later in this chapter but it is necessary at this point to state that, given the way UGBs were established, any fears regarding dislocations in land values inside those boundaries were exaggerated but not totally unfounded (Ketcham and Siegel 1991).

The acceptance of housing as a state planning goal

State Planning Goal 10 requires local jurisdictions, through their planning efforts, "to provide for housing needs of citizens of the state." While it calls for an inventory of buildable land for residential use it goes beyond that to introduce social equity as a central objective. This comes out clearly in the requirement that "plans shall encourage the availability of adequate numbers of housing units at price ranges and rent levels which are commensurate with the financial capabilities of Oregon households." Similarly, Goal 14, while concentrating on the need for an orderly urbanization process, established the link between growth management and housing through a recognition of the relationship between buildable land and housing affordability. In principle the two goals are linked through an implicit common interest in enhancing the availability of housing and in maintaining the smooth operation of the housing market. Beyond that general interest, policy focus and programmatic emphasis emerged along diverging tracks with Goal 10 providing the central framework for the state housing policy. It should be noted, however, that Oregon's land use goals are not mutually exclusive and are implementable only through local plans that are required to be in compliance (LCDC 1985).

Like any new phenomenon, SB 100 and the planning process it produced were welcomed by some and dreaded by others. with skeptics taking a wait-and-see attitude. The bill pleased environmentalists and generated considerable apprehension among developers, but housing activists remained on the sideline, unclear as to how the bill affected their concerns, especially those related to social equity in the housing market. Vocal concerns about the bill's negative impacts on the operations of the housing market came mostly from developers and home builders whose objections were directed more towards Goal 14 rather than Goal 10. However, in the absence of clear evidence that the two goals threaten the interests of builders, organized opposition failed to materialize. Opposition to the housing objectives of the Oregon land use planning process was muted at best. At the risk of appearing cynical one could argue, therefore, that whatever contributed to the indifference of those advocating liberal housing policies may have also reduced the level of anxiety among opponents of land use planning and regulations, at least in so far as housing issues were concerned. One explanation, perhaps, could be found in the fact that Oregon's planning process did not require impact statements and did not call for the preparation of a statewide plan (Sullivan 1990). The goals were viewed as guidelines to be utilized by local governments in preparing their own plans which, if past precedents were any clue, were unlikely to address questions of housing equity or cause undue hardship to developers. At least that was the general interpretation. In reality both sides underestimated the intent of the legislature and the determination of LCDC to uphold the law and ensure compliance. This scenario, while difficult to prove, receives some credibility from the fact that in the early years most of the challenges to the housing goal came from local governments that, intentionally or unintentionally, produced the majority of test cases that defined the legal limits of the legislation.

The initial skepticism by which low-income housing advocates approached the Oregon planning process could also be attributed to perceived ambiguities in the way Goal 10 was to be implemented. In reality, there was nothing ambiguous in the guidelines established by LCDC to assist local government in planning for and implementing the housing goal. The objectives and the means to achieve them were clear, but they were stated in language that was not much different from that commonly used in federal legislation and local comprehensive plans. Such language was symbolic and lacked any discernible impact on the availability of affordable housing. In its early years the Oregon planning process was accepted by most as a positive contribution to urban and regional planning subject to validation of the state's true intent regarding implementation. While this statement is applicable to the entire process, it is particularly true in regard to the housing goal because of the high degree of cynicism that developed over the years regarding the place of housing in the overall comprehensive planning process. Further legislative actions, court decisions, rulings by the Land Use Board of Appeals (LUBA), and LCDC directives provided the needed tests that prepared the ground work for today's general acceptance of the state's role in guiding and monitoring local actions that affect the availability of affordable housing. Of all these actions the court decisions and LUBA rulings are the most significant.

Legal challenge and affirmation

Oregon's housing goal was never challenged in its entirety. This fact lends support to the argument regarding the lack of serious opposition to the spirit embodied in Goal 10 and to the housing objectives of Goal 14. However, numerous challenges were filed against or in support of specific provisions. Throughout these challenges the courts and LCDC held firmly to the view that Goal 10 "imposes an affirmative duty on local governments to provide reasonable opportunity for the private sector to supply housing units at prices and rents within the financial capabilities of current and *prospective* area residents" [emphasis added] (CLE 3-14). In subsequent actions the legislature confirmed that interpretation, thus placing the burden of proof on governments. According to Sullivan the notion of affirmative duties was based on the principles of "fair share," "least cost," and "the St. Helens Policy" (Sullivan 1990).

The first two principles were enunciated by the New Jersey Supreme Court in *So. Burlington Cty. NAACP v. Tp. of Mt. Laurel*, 336 A2d 713, 732-733 (NJ 1973) and *Oakwood at Madison, Inc. v. Tp. of Madison*, 371 A2d 1192, 1207-1208 (NJ 1977) (CLE 3-14). The St. Helens Policy is, in reality, a policy statement adopted by LCDC in 1979 to clarify an earlier interpretation issued during the review of the City of St. Helens comprehensive plan. The principal purpose of the policy as stated is to ensure the provision of adequate numbers of "needed housing types" in a community at least cost. The policy required local jurisdictions to permit, in a zone or zones where buildable land is sufficient, particular housing types that are needed to satisfy the desirable cost and rent mix. In other words, the policy prevented communities from using restrictive zoning regulations to bypass fair share and least cost.

One of the earliest and perhaps the most significant of cases to test the applicability of the fair share principle is *Seaman v. City of Durham*. (1 LCDC 283 1978). In that case the City of Durham amended the definition of the A-1 zone in its comprehensive plan to reduce, by half, the permitted density for single-family homes, duplexes, and multiplexes. Furthermore, the City zoned all remaining residential areas into single-family zoning with a minimum lot size of 15,000 square feet. The ordinance was invalidated on the grounds that it failed to account for the low-cost housing needs of its residents and, in support of fair share, those of the region (CLE 3-15). The message of upholding fair share was clear: Goal 10 protects the housing interests of the all households in a given region and cannot be applied selectively on the basis of local interests. The ruling against the City of Durham is also important because of its implicit affirmation of the least-cost principle that requires localities to adjust their regulation so as to render feasible the construction of least-cost housing.

Attempts by local governments to amend their charters in order to circumvent the intent of the state's housing goal were nullified by actions of the state legislature (*State of Oregon v. City of Forest Grove*, 9 Or LUBA 92 1983). In 1983 ORS 197.295-197.313 was amended to recognize government-assisted housing as a separate needed housing type for which provisions should be made. On the matter of amending a charter to exclude certain types of housing, ORS 197.312 states that no government "may by charter prohibit from *all* residential zones attached or detached single family housing, multiple family housing for both owner and renter occupancy or manufactured homes (emphasis added)." The statute goes on to prevent governments from prohibiting or imposing additional approval standards on government-assisted housing.

In another ruling that contributed to closing loopholes and gave a boost to controversial types of low-income housing, the court decided that governments cannot impose special conditions with the intent of excluding certain types of low-income housing (*City of Hillsboro v. Housing Devel. Corp.*, 61 Or App 484, 657 P2d 726 1983). In that case the City of Hillsboro interpreted its zoning ordinance to require conditional use permits for migrant housing projects even though these projects do not deviate from the city's definition of multifamily housing. The city's interpretation was based on the claim that the character of migrant housing required a special use permit but the court disagreed, thus ruling out character as a criterion as long as the project complied with established definitions. In different rulings the same protection was extended to mobile homes and manufactured housing. Given the importance of these housing types to meeting the needs of low-income households, such actions served to validate the housing goal's intention to strengthen social equality.

A year later the court clarified fair share and least cost intents as they apply to small communities with populations below 2,500. In *City of Happy Valley v. LCDC* (66 Or App 795, 799-801, 677 P2d 43, aff'd as modified, 66 Or App 803, rev. denied, 297 Or 82, 1984) the Court of Appeals ruled that, while such communities are exempt from providing housing mix by type, they are nevertheless required to permit needed housing at particular price ranges and rent levels. In doing so, the court upheld fair share as a general principle but allowed smaller communities to follow a less stringent interpretation. That interpretation gave LCDC the option of requiring such housing types as a condition of acknowledgment.

Another aspect of Goal 10 that was enforced by LCDC and upheld by the courts is the requirement that local government prepare an inventory of "buildable land." Communities are required to develop such inventories for residential development needs in their jurisdiction over a twenty-year period. The definition of buildable lands excluded all lands outside urban growth boundaries (Ragatz 1979) and was interpreted to include all vacant land inside those boundaries available or suitable for residential development. That interpretation was affirmed by LCDC in several cases including those of cities inside metropolitan areas such as Lake Oswego and outside such as La Grande and Newport. In a case involving the City of Redmond, the definition was narrowed to exclude vacant land that is not serviceable in the planning period and cannot be serviced in the long run. In enforcing compliance LCDC required that inventories include a breakdown by type and allowed communities to evaluate in-fill potential on oversized lots. In some cases, including the City of Portland, the commission ruled that in-fill and redevelopment may be used to meet part of the identified housing needs (CLE 3-18). It was through such interpretations that LCDC reached past vacant land in the urban fringe to encourage local government to reexamine residential development policies in established neighborhoods. Obviously, the social significance of such an interpretation cannot be underestimated. In fact, it was this aspect of LCDC housing policy that left the greatest impact on Portland, the state's largest city.

In dealing with the housing needs assessment, LCDC dispelled any doubts about the social significance of the state's housing goal and the seriousness with which it is taken by the commission as well as by the legislature and the courts. The needs assessment itself was viewed as a policy decision, with all the flexibility that entails. But it must be based on factual information, defined to include all data necessary for a comprehensive market analysis study. However, enforcement has been carried out on the basis of liberal interpretations that accounted for differences between communities, including their abilities to conduct sophisticated studies and analysis. As a minimum, all are required to provide information necessary to determine housing types and densities appropriate to encourage housing at affordable costs (CLE 3-22).

In another application of the housing needs assessment provisions, LCDC ruled and the courts agreed that regional needs must be accounted for. As a result, housing needs in the Portland and Eugene metropolitan areas have been determined on regional bases with all communities, as already discussed above, held responsible for their fair share. In the Portland metropolitan area this led to the adoption of the Metropolitan Housing Rule (MHR) that became as significant as Goal 10 itself in shaping housing policy in the area under the jurisdiction of the Metropolitan Service District (Metro). In fact, it is this rule, adopted in 1981, that is credited with most of the positive developments that took place in the Portland housing market during the last ten years. Praise for that rule has come both from those who in earlier years supported and those who opposed SB 100 (Ketcham and Siegel 1991).

The Metropolitan Housing Rule

The Metropolitan Housing Rule (MHR) was adopted as Division 7 of chapter 660 of the Oregon Administrative Rules. Its stated purpose is "to assure opportunity for the provision of adequate numbers of needed housing units and the efficient use of land within the Metropolitan Portland (Metro) urban growth boundary, to provide greater certainty in the development process and so to reduce housing costs." This rule, while intended as an affirmation of the commission's view of the importance of addressing the metropolitan housing market as a single entity, created an environment supportive of regional planning for housing needs. MHR did not alter LCDC's primary function as a regulatory agency, but moved it one step closer to planning through the application of variable standards designed to account for differences between local communities and to produce a variety of residential patterns. The rule required Metro to include in its periodic reviews of its UGB a determination regarding the sufficiency of buildable land to satisfy projected housing needs for the region. It also called on Metro to ensure that regional housing needs are met through coordinated comprehensive plans. In calling for such coordination, LCDC came as close as it can to mandating a lesser form of regional development planning.

Coordination was also encouraged on a different level. Through the enactment of the metropolitan housing rule, LCDC brought about close coordination in the implementation of Goals 10 (housing) and 14 (urbanization). Implementation of the rule was intended to achieve the basic objectives of the housing goal: providing an appropriate housing mix and enhancing housing affordability. It was also designed to contribute to the success of the Metro UGB by mandating minimum average densities and housing mixes to maintain the availability of buildable lands throughout the planning period. In theory, therefore, it can be argued that MHR, which came eight years after the adoption of SB 100, integrated housing with all its social concerns into the mainstream of the Oregon land use planning process. This conclusion finds support in a 1991 study sponsored by 1000 Friends of Oregon and the Home Builders Association of Metropolitan Portland. That study, which covered the five-year period 1985-89, determined, among its many findings, that land use planning was a major contributor to the provision of housing needs of the region and that the "region's *pro-housing* policies have helped to manage regional growth while promoting affordable housing" [emphasis added] (Ketcham and Siegel 1991, p. 68).

The rule, which has been given considerable credit for the success of Goals 10 and 14 in the Portland area, relies heavily on a set of average residential densities that became known as the 6/8/10 formula. The density measures stipulated by this formula (ORS 600-07-035) apply to the areas of Clackamas, Multnomah, and Washington counties that are inside the Metro UGB. As a measure of realism, jurisdictions that do not provide the opportunity for "at least 50 percent of new residential units to be attached single-family housing or multiple-family housing" were excluded from the list of jurisdictions required to comply with the average density standards. This exclusion applied to a handful of small cities that in 1977 had less than 50 acres of buildable land. The region's six largest cities as well as its urban county (Multnomah) are required to provide for an overall density of ten or more dwelling units per net buildable acre. All these were, at the time, jurisdictions with regionally coordinated projected populations of 50,000 or more. In other words, the great majority of the region's projected population is to be accommodated in areas with relatively high residential densities. A reduced average density of eight dwelling units per acre applies to most of the remaining areas within the UGB, with the exception of five small communities with projected populations of 8,000 each, for which density was further reduced to six dwelling units per acre.

One of the positive elements in MHR has been the flexibility offered to local jurisdictions that opt for innovative variations to the density and housing-mix guidelines. New construction could be exempted from the density standards if an alternative mix is accepted. The rule requires that densities of single and multiple housing must equal or exceed the average densities for each housing type that existed in the plan at the time of the original acknowledgment. MHR, therefore, is not a rigid regulatory tool that carries the risk of suppressing innovation or creating hardships for developers or local communities. It encouraged local communities to engage in creative thinking and, to some degree, rewarded those who did so.

MHR's contribution to the Oregon land use planning process has been considerable, even though it may have not been by design. The rule was developed to regulate residential development in the state's largest urban region in accordance with the objectives of Goal 10. As already indicated, the process and its effectiveness was not universally acknowledged when it was first introduced. The greatest skepticism was directed to the notion of delineating urban growth boundaries. While some felt that the boundaries were too big, many others feared that they would likely stifle orderly development on the inside and become ineffective on the outside. The results produced by implementing the metropolitan rule leave the question of the boundary's size unanswered, refute any arguments about negative impacts on the housing market, and raise serious questions about the effectiveness of the UGB in areas outside its limits.

Recent studies that evaluated the impact of Goal 10 and the effectiveness of Goal 14 point to the disappointing fact that, except in the Portland area, large percentages of residential development are occurring outside the approved UGBs. A study prepared for the Department of Land Conservation and Development revealed that in the Bend area (one of the fastest growing communities in the state) 57 percent of residential development occurred outside the UGB (ECO Northwest 1991, p. 7). Among the other three cases covered by the study, only Portland fell below 5 percent. In Brookings, on the coast, and the Medford metropolitan area, at the southern end of the I-5 urbanized corridor, the percentages of development outside the UGB were 37 and 24 respectively. The study also revealed that the Portland area had the lowest percentage of single-family homes built outside the UGB (9 percent compared to 63 percent in Bend). Of greater significance perhaps is the study's finding that "more single family units (2702) were developed outside the UGB in Deschutes County [where Bend is located] than in the three counties of the Portland metropolitan area," despite the considerable difference in size.

The issues raised by residential development outside the urban growth boundaries are significant as indicators of uncertainties in the concept of growth management as applied in Oregon and deserve serious examination. Do they signal failures or are they the product of normal slippage? The answers to these questions are likely to be different depending on which area is examined. In the context of the MHR, it is noteworthy that development outside the Portland UGB is not as extensive as in other areas and that the impact of outside development on the housing market is, therefore, marginal at best. As time passes, the situation is likely to change and, at that point, what is happening outside the UGBs could have negative implications on the availability of affordable housing. For example, an emerging belt of very low-density residential areas is certain to pose a formidable challenge to the future urban form of a growing region. This danger is greatest in Clackamas County. For the time being, however, it seems that the Portland area has had greater success in meeting the objectives of Goal 14 than most other areas in the state. It is difficult to prove whether this is the result of MHR, given the enormous differences in size and demographic and socioeconomic conditions that differentiate the Portland area from the rest of the state. A reasonable test can, however, be developed using density and housing mix, two indicators that are intrinsically tied to the MHR objectives.

Density. As indicated earlier, MHR regulates densities in the region using a 6/8/10 formula that was designed not only to meet the objectives of Goal 10 but also to produce a development pattern that complies with the urban growth rules of Goal 14. Shortly after the rule came into effect 1000 Friends of Oregon attempted to illustrate the benefits that would accrue to consumers as a result of the proposed density guidelines (1000 Friends 1982). That study estimated that, in 1978, 82 percent of all vacant land within the Portland metropolitan region zoned single-family residential was zoned for lots 10,000 sq. ft. or more, the average being 12,800 sq. ft. By 1989 single-family homes inside the UGB were being built at an average density of almost 5 units per acre. This translates into an average lot size of about 8,800 sq. ft., not much above the 8,200 sq. ft. stipulated by MHR (Ketcham and Siegel 1991, pp. 29-39). Because multifamily units were being built at an average density of more than nineteen units per acre, the overall 1989 density in the region was nine units per acre, substantially more than the 6.23 figure used in determining the size of the Metro UGB (ECO Northwest, 1991, p. 22).

Within the three density zones defined by MHR, compliance was reasonably good in the areas with higher density requirements but fell substantially below the targeted level in the few areas with a low density requirement. With this exception, the overall targets were generally met, though some differences in degree of compliance exist between local jurisdictions. Areas with a six-units-per-acre standard actually achieved 3.09 units per acre. Areas aiming for ten units per acre fell slightly short (9.58), and those aiming for eight units per acre exceeded the targeted level (8.41). The situation looks somewhat different when only new development is accounted for. In 1991, the 1000 Friends/Home Builders Association study (Ketcham and Siegel 1991) found in the Portland area "local government approved residential development at 79 percent of the maximum densities allowed in their approved comprehensive plans." In other words, even though MHR targets were generally met, densities could still be increased if communities stay closer to their plans. It is important to note, however, that Portland shows a significantly higher level of compliance than other areas in the state (ECO Northwest 1991, p. 21).

Housing mix. Among the four cases studied by ECO Northwest, multiple-family residential development was highest in the Portland metropolitan area (54 percent of total). By comparison multiple-family development represented 34 percent of the total in Brookings and only 15 percent in Medford. The explanation for Portland's higher proportion of multifamily units lies in the rezoning caused by the application of MHR. In 1982, 1000 Friends of Oregon estimated that zoning changes increased by almost 400 percent the amount of land available for multi-family residential development and predicted that a sharp increase in the construction of such units was imminent. Available figures suggest that this conclusion was on target. In fact, the increase in multifamily construction was such that the 50:50 mix called for in the MHR was exceeded in all three metropolitan counties. Interestingly enough, it is Multnomah County that shows the lowest ratio of multifamily to single-family development (52:48 compared to 56:44 in Clackmas County). Compliance with the stipulated housing mix is not uniform. During the last five years of the 1980s communities where the housing mix exceeded 65:35 were all in the suburban belt, with Beaverton, Forest Grove, and Oregon City showing a ratio of about 3:1. Portland, the largest city in the region and its central core, with a ratio of 48:52 is actually one of the few communities that failed to meet the MHR target.

Obviously the changes that occurred in development patterns and trends in the Portland area cannot be attributed to MHR alone. Indeed, there is a strong cause and effect relationship between densities, housing mixes, and the success of growth management within UGBs. Similar relationships exist among housing trends, demographic changes, employment trends, and other social indicators. However, and regardless of any externalities, the end result of what happened in the Portland area lends credibility to the conclusion reached by 1000 Friends of Oregon and the Home Builders Association that MHR has been a very effective tool (Ketcham and Siegel 1991). What cannot be easily verified is the extent to which housing affordability has been enhanced by the enactment of MHR.

Local response

As indicated earlier in this chapter, central cities were ahead of the states in recognizing the need for coordinated action in the area of housing. Most, however, utilized the federal connection and concentrated their efforts on government-supported or -assisted housing and in the process failed to develop areawide comprehensive housing policies. The kind of issues addressed by the Oregon state housing and urbanization goals were rarely incorporated in comprehensive plans. This does not mean that the problem did not exist, nor does it mean that people were not aware of it. Quite to the contrary, housing literature was rich on the subject of affordable housing and the shortcomings of prevailing policies and attitudes in meeting its objectives. What was missing was an acceptable solution and a receptive political climate. SB 100 provided that climate and encouraged local government in the state to move beyond the limitations of federal programs and begin searching for home-grown solutions.

In Portland, the foundations for a more comprehensive housing policy preceded SB 100. The Downtown Plan, developed in 1972 and released a few months before SB 100 was adopted, is known not only because of its vision for a healthier downtown but also for its enlightened housing policy. A major goal of the plan was "to increase the supply of downtown housing for all income groups" (Downtown Committee 1972, pp. 31-36). The plan advocated the use of what were then traditional approaches to encourage moderately priced housing. These included controlling land values in urban renewal projects, greater involvement by the Portland Housing Authority, and incentives for higher densities. Its guidelines, however, included some that became precursors for significant actions that, several years later, altered drastically the city's approach to housing policy. Among these guidelines, two are particularly important: a firm and liberal replacement policy that sought to maintain the availability of affordable housing and a call for a citywide coordination effort to engage all city agencies as well as the private sector in housing planning.

By the late 1970s the City of Portland was firmly engaged in discussing socially sensitive housing policies through two citizens' advisory committees. One of these addressed social policy and the other dealt with housing. The Housing Advisory Committee remains active today and has been a major contributor to ideas aimed at meeting MHR objectives. How much of the change of attitude in the city since 1972 is due either to SB 100 or to MHR is difficult to assess. It should be remembered that 1973, the year SB 100 was adopted, is also the year Congress enacted the Housing and Community Development Act that tied federal support to the development of housing plans. At best, therefore, one can argue that SB 100, if not the only catalyst for change, provided the means for enforcement and the incentives needed to more actively engage local government. Active engagement in housing planning became a requirement for all jurisdictions in the state but MHR added greater incentives for those in the Portland area.

A recent study released by Metro in 1991 found that all three metropolitan counties were seriously engaged in examining countywide housing needs (Metropolitan Service District 1991). Multnomah County in cooperation with the cities of Portland and Gresham is moving ahead with the implementation of a Comprehensive Housing Affordability Strategy (CHAS). The central objective of CHAS is "to adopt and implement an innovative plan for housing opportunities and support services emphasizing affordable housing for no-, low-, and moderate-income people through maximizing resources and coordination among all levels of government and the private sector" (Multnomah County 1991, p. 3). The report adopted by the county and the two cities in December 1991 is an elaborate statement by the three governments that goes much beyond the requirements of MHR. There is no doubt, however, that this strategy, as well as policies on housing in-fill adopted by the City of Portland, reflect an awareness that more needs to be done to comply with MHR housing mix requirements[13] (Metropolitan Service District 1991, p. 28). As a result the influence of Goal 10 and MHR cannot be discounted.

Washington County, the region's second-largest and fastest growing county, has been operating with a 1983 Comprehensive Frame Work Plan that includes four housing policies that deal with housing affordability, housing choice and availability, housing conditions, and housing discrimination. The plan espouses several strategies that are designed to encourage in-fill with "compatible development," to review design and development to reduce cost, and to increase densities in unincorporated areas.

Positive response to Goal 10 and MHR is not limited to the larger jurisdictions. Smaller communities in the Metro region were receptive to the fair share/least cost requirements. Some went only as far as needed to obtain LCDC compliance acknowledgment but many were forthcoming in their search for ways to comply without seriously impacting their perceived values and character. A significant example is the City of Happy Valley, a community incorporated with an eye on maintaining its rural character. The city's comprehensive plan as acknowledged by LCDC allows for the building of secondary residential units (accessory apartments) on already developed existing single-family lots. The idea is to provide for increased densities through small single-family rental homes suitable for single individuals and elderly couples, thus meeting some of the requirements of the housing mix and density without serious deviation from the prevailing community character.

At the regional level, Metro has been actively pursuing the development of housing plans since 1979 when an areawide Housing Opportunity

Plan was approved and supported by almost all jurisdictions including some that are not inside Metro's territory (Metropolitan Service District 1991, p. 32). The plan attempted to assess housing needs by jurisdiction and incorporated a model for the distribution of federal funds for assisted housing. The availability of federal funds was a key element for the success of the plan. As a result, the cuts enacted by the Reagan administration led to its demise. More recently, Metro completed a Regional Urban Growth Goals and Objectives statement (RUGGO). Its housing policy addressed issues similar to those addressed in the Washington County Plan. The housing objective states that "there shall be a range of housing types available inside the UGB for rent or purchase at costs in balance with the range of household incomes in the region." The influence of Goal 10 and MHR is clear but Metro goes further in requiring that "housing should be located in proximity to major activity centers and regional transportation system."

In summary, local responses in the Portland area to the recommendations of Goals 10 and 14 as well as to the Metropolitan Housing Rule have been generally positive and supportive. Like all regulations, MHR did generate resentment and friction, but the flexibility by which LCDC approached the process of implementation avoided outright hostility and eventually succeeded in attaining many of the desired outcomes. In many cases MHR also encouraged local government to seek independence in their own policies; some are actually more supportive of higher densities and affordable housing than LCDC mandated. Discussion here has been limited to local jurisdictions in the Portland area because of the significance of MHR in forcing them to become more active in housing issues. This does not imply that other jurisdictions in the state were less receptive to housing needs and problems. However, the LCDC Growth Management Study and the cases it examined suggest that outside the Portland area, and in the absence of regulations comparable to MHR, compliance with the density and housing mix objectives of Goals 10 and 14 was less successful.

Growth Management and Housing Affordability

Housing affordability has been given a central place among the goals developed to implement SB 100. It was also a major concern of those who, in the absence of precedents and convincing scientific evidence, feared for serious dislocations in the residential land market. As indicated earlier, many of these fears were unwarranted. In fact, the Portland UGB does not seem to have created any imbalances in the land market. This conclusion, reached by the 1000 Friends/Home Builders' study, applies to both values and land availability (Ketcham and Siegel 1991, pp. 40-46). The study based its conclusions on the fact that MHR led to a reduction in lot sizes and to an increase in the amount of land available for multifamily housing.[13] Obviously, the inherent assumption is that both trends should contribute to the availability of affordable housing. Using 1989 rent, cost, and income data the authors reached two general conclusions: 1) "Goal 10 implementation has helped to mitigate shortages of affordable housing by allowing development of a greater proportion of multiple family units"; and 2) "MHR implementation has helped to mitigate shortages in affordable housing by allowing development of single family housing on smaller less costly lots." The increase in the number of multi-family units and the low vacancy rates reported during the study period were used, correctly, to assert that the demand for such units is high and that more development of such housing is still desirable. The same data were used by the authors and borrowed by DLCD to criticize the Portland area for having approved development at 79 percent of the maximum allowed densities, thus threatening the future availability of affordable housing (ECO Northwest 1991, p. 9). Both studies called for adjustment to MHR standards to encourage downzoning so as to maintain the availability of land for affordable housing. In general, therefore, there seems to be general agreement that the growth in the number of multifamily units represents a growth in affordable housing types, but does not necessarily imply that housing has become more affordable.

In reaching the second conclusion the 1000 Friends study used changes in per capita income and in the median price of single-family homes. Between 1979 and 1990 per capita personal income grew by more than 73 percent while median prices increased by about 30 percent. Unfortunately, this type of analysis proves very little. For example, careful examination of the annual variations provided in table D-1 of the study reveals that the gap between growth in personal income and median prices is due to the severe depression that occurred in the Portland housing market during 1982-85 (Ketcham and Siegel, 1991 Appendix D, pp. A30-A42). Prices actually declined and did not reach 1982 levels again until 1988. It is difficult to attribute this decline to Goal 10 or MHR. For the period 1985-90 median home prices increased at a rate faster than that of personal income (26.8 percent compared to 24.1 percent). Again it is difficult, and perhaps unfair, to draw any connection between growth management and recent increases in home prices. Table D-4 of the same study suggests that between 1985 and 1990 the average rent of a two-bedroom apartment increased by 43 percent, considerably more than the growth in personal income. Using these simple analyses one can easily conclude that, in reality, housing has become less affordable regardless of the changes brought about by MHR. That was the conclusion reached by the ECO Northwest study and incorporated by DLCD in their final report. However, a complex issue like affordability cannot be addressed by simple comparisons.

Affordability is a relative phenomenon. The increase in the number of affordable types cannot by itself indicate that housing today has become more affordable than before MHR was enacted. To reach a firm conclusion, changes in housing rents must be compared to many variables including household incomes. Using the same logic, one can also argue that judging Goal 10 and MHR's success in meeting their affordability objectives is difficult without knowledge of what would have happened in their absence. No good purpose will be served in trying to speculate. Evaluation of the state housing goal's performance on affordability cannot be achieved without comparative analysis using national trends to control for external changes. In the absence of such comparative studies, let us look at a different question: did Oregon's land use program produce market changes that may have negatively affected affordability?

One possible answer was provided by Metro, which compared Portland to a group of similar western cities. The emerging evidence suggests that the UGB did not increase pressure on the land market (Metropolitan Service District 1991, Table 13). The finding is based on the fact that between 1980 and 1990 the average price of a 10,000-sq ft. improved single-family lot in Portland grew at an annual compounded rate of 3.6 percent compared to a western average of 4.9 percent. Only Tacoma had a lower rate of increase. The 1990 average price of $31,250 placed Portland sixth among the nine cities studied, with Tacoma, Salt Lake City, and Phoenix as the three cities with lower prices. In 1975 Portland had tied Phoenix for fifth place. Using such information it is easy to conclude that growth management in the Portland area did not result in abnormal increases in land prices.

The 1000 Friends/HBA study (Ketcham and Siegel 1991) also provides comparative income and cost statistics for 23 metropolitan areas. In thirteen of those areas the 1990 income needed to qualify for the medium-priced home was less than the median household income but only in three of them did the difference exceed that reported for Portland. Affordability indices developed for the same 23 areas placed Portland fourth after Houston, Detroit, and Minneapolis/St Paul as an affordable housing market. The same data suggest that affordability declined between 1987 and 1990, though evidence suggests that the decline is rooted in national trends rather than in local phenomena. Given these findings, it is again safe to assume that if growth management did not enhance housing affordability, it also did not diminish it. In fact, the growth in the availability of rental housing, which is a direct product of growth management, leaves Portland in a stronger position to face the challenge of the next few years. This challenge as summarized in the 1991 *State of the Nation's Housing* lies primarily in the anticipated shortages in the rental market (Joint Center for Housing 1991, p. 23).

Conclusion

The review conducted in this chapter suggests that Oregon's experience with incorporating housing policy in the statewide land use planning process provides useful lessons for others attempting to use the same or similar approaches. The approach has been generally successful when regulations and guidelines went all the way in defining the expected outcome. Almost all success stories are in the Portland metropolitan area, where a housing rule was adopted defining the parameters within which local jurisdictions are supposed to operate. It is also clear that developers and builders were not the ones who rushed to challenge the strict density and housing mix requirement of MHR. Rather, most of the challenges came from local governments interested in minimizing change in community character. Another important lesson stems from the flexibility that LCDC applied to enforcement. The result has been less than uniform compliance levels but greater levels of acceptance. As the process evolves and success rates increase, the credit should be given to the enlightened approach that the commission followed. But it is still too early to reach a firm conclusion on the extent of success or failure in the application of the housing policy. Simple correlations suggest that housing affordability declined and congestion increased (ECO Northwest 1991, p. 27). On the matter of affordability, however, it is perhaps better to reverse the question in order to ask whether growth management did any harm to the housing market. The conclusion reached in this chapter is that there is no evidence that, with growth management and strict housing rules, the Portland area is in any worse position than similar areas in other parts of the country. This is by itself an adequate measure of success,z given the other benefits of growth management and the fact that a system for guiding future housing policy is already in place, is working, and can only improve with the passage of time.

Notes

1. In his book *The City in History* Lewis Mumford attributes the great plague during the Peloponnesian War to the lack of sanitary conditions which, when coupled with the overcrowding produced by the war, created suicidal conditions in Athens. He also establishes a correlation between rising infant mortality and deteriorating housing conditions. As an example, in New York City "the mortality rate for infants in 1810 was between 120 and 145 per thousand live births; it rose to 180 per thousand by 1850, 220 in 1860, and 240 in 1870."

2. In *Land Use Planning*, Charles Haar points to some earlier regulations that include an ordinance adopted by the City of Philadelphia in 1796 to prohibit the erection of wooden buildings in a specifically described area of the city. That ordinance was based on a 1795 enabling legislation and as a result it survived challenges in the courts but other cities that attempted similar regulations were not as successful. On the federal level early regulations dealt with the disposition of public lands and homesteads.

3 Kent defined the General Plan as "the official statement of a municipal legislative body that sets forth its major policies concerning desirable future physical development; . . . the document must include a single, unified general physical design for the community, and it must attempt to clarify the relationship between physical-development policies and social and economic goals." Based on the contents of the plan and on the discussion of its elements that followed that definition, social goals were clearly perceived as inputs to the planning process and not necessarily central elements (Kent 1964, pp. 18-26).

4. It is interesting to note that in addressing the question of future housing needs Chapin's assumptions dealt with three main categories: changes in

household size, losses in the existing stock, and changes in the vacancy rate. Missing in that approach was the issue of affordability and planners' social responsibility with regards to the future provision of housing for all.

5. The 1949 Housing Act, better known as the Decent Home Act, represented the first successful attempt by Congress to spell out national commitment to provide decent housing for all who need it. However, it was also the act that introduced urban renewal, which destroyed more low-income housing than it provided (Anderson 1964). While a landmark, the act never produced the desired results. A decent home for all remains an illusive goal and, as Martin Anderson's book *The Federal Bulldozer* illustrated, urban renewal became shrouded in controversy that eventually reduced its effectiveness as a major force in shaping the future of our cities.

6. Local housing authorities and redevelopment agencies regardless of their local mission and identity were viewed as creatures of the federal government and an extension of its influence in the cities. While some of these perceptions remain with us today, the diminishing role of Washington is slowly altering the overall picture and the balance of power between home grown institutions and those created as a result of federal initiatives.

7. McCall left his mark on Oregon politics through his nationally publicized campaign to discourage growth in the state. "Visit but do not stay" became a slogan that symbolized his crusade on behalf of environmental protection. Such arguments against growth came very close to stifling economic development in the state and confused the issue. It was not growth per se that was the villain; rather it was the way it was taking place. However, McCall's real contribution was the creation of a political climate that embraced land use and environmental planning on regional and state levels. The Willamette Greenway plan developed in the early 1970s is a good example of McCall's contributions. That plan was a catalyst that revolutionized the public attitude toward land use planning.

8. By the end of the seventies the shift away from physical planning reached extreme levels and the overall impact on the profession became controversial. The nature of the controversy is not relevant to the questions at hand. It is the fact that the shift in emphasis took place that concerns us here.

9. The Green Acres program of the State of New Jersey, developed in the early sixties, is a good and early example of a statewide attempt to deal with urban sprawl. Concerns with the program dealt with its economics especially potential costs but it fell short of being a mandate for statewide planning.

10. A good example, perhaps, is New York State Urban Development Corporation which, regardless of its emphasis on specific projects and finance programs, gave New York a different image than most states.

11. That conclusion is justifiable only because the bill unleashed a chain of events that led to the adoption of more comprehensive legislation that placed housing among the central concerns that the state must address.

12. In 1990 testimony, Edward Sullivan listed seven of the nineteen goals as being significant in one way or another to the quest for affordable housing. As it happened these goals are the ones central to the overall planning and development policy. Concerns covered by these goals included housing, transportation, facility location, energy conservation, growth management, and land use planning. Since the planning process is expected to draw a balance between all elements of the urban system, Sullivan is technically correct in his interpretation and as such he was only restating the nature of comprehensive planning. However, his testimony is introduced here because it represents views on the centrality of housing to the Oregon planning process that are shared by many housing activists.
13. Of the three governments involved in the development of CHAS, only Gresham exceeded MHR's housing mix target. Portland fell 2 percentage points short and the unincorporated areas of the county were substantially below target (Ketcham and Siegel 1991, p. 26, Table 2).

References

Anderson, Martin. 1964. *The Federal Bulldozer.* Cambridge, MA: M.I.T Press.

Basset, Edward M. 1938. *The Master Plan.* New York: Russell Sage Foundation.

Bettman, Alfred. 1929. "City Planning Legislation." In *City Planning: A Series of Papers Presenting the Essential Elements of a City Plan,* edited by John Nolen. New York: Appelton and Co.

Chapin, F. Stuart. 1957. *Urban Land Use Planning.* New York: Harper and Brothers.

Davidoff, Paul, and Thomas Reiner. 1962. "A Choice Theory of Planning." *Journal of the American Institute of Planners* 28, 2: 103-15.

Downtown Committee. 1972. *Planning Guidelines: Portland Downtown Plan.* Portland: City of Portland.

ECO Northwest. 1991. *Urban Growth Management Study: Case Studies Report.* Prepared for Oregon Department of Land Conservation and Development.

Gallion, Arthur B. 1950. *The Urban Pattern.* Princeton, NJ: D. Van Nostrand Co.

Geddes, Patrick. 1968. *Cities in Evolution.* New York: Howard Fertig, Inc. First published 1915.

Haar, Charles. 1977. *Land Use Planning: A Casebook on the Use Misuse and Reuse of Urban Land.* Boston: Little Brown and Company.

Howard, Ebenezer. 1902. *Garden Cities of Tomorrow.* London: S. Sonnenschein and Co., Ltd.

Joint Center For Housing. 1991. *The State of the Nation's Housing 1991.* Cambridge: Harvard University.

Kent, T.J. Jr. 1964. *The Urban General Plan.* San Francisco: Chandler Publishing Co.

Ketcham, Paul and Scott Siegel. 1991. *Managing Growth to Promote Affordable Housing: Revisiting Oregon's Goal 10.* Portland: 1000 Friends of Oregon and the Home Builders Association of Metropolitan Portland.

Klaus. Susan L. 1991. "Efficiency, Economy, Beauty. The City Planning Reports of Frederick Law Olmsted, Jr. 1905-1915." *Journal of the American Planning Association* 57, 4: 456-470.

Land Conservation and Development Commission. 1985. *Oregon's Statewide Planning Goals.* Salem: Oregon Department of Land Conservation and Development.

McHarg, Ian. 1969. *Design with Nature.* New York: Natural History Press.

Metropolitan Service District. 1991. *Housing Issues Report.* Portland: Planning and Development Department, Metropolitan Service District.

Multnomah County Housing Advisory Committee. 1991. *CHAS: A County-Wide Housing Affordability Strategy, 1991-1996.* Portland: Multnomah County.

Mumford. Lewis. 1961. *The City in History.* New York: Harcourt, Brace, and World.

Nelson, Arthur C. 1984. *Evaluating Urban Containment Programs.* Unpublished Ph.D. Dissertation, Portland State University.

1000 Friends of Oregon. 1982. *Responding to the Market Place: How Oregon's Land Use Program Has Benefitted Housing Consumers in Metropolitan Portland.* Portland: 1000 Friends of Oregon.

Oregon State Bar, Committee on Continuing Legal Education (CLE). 1982, 1988. *Land Use; Volume 1*, and 1988 supplement.

Oregon Legislative Assembly. 1969. Corrected Senate Bill 10 (1969 Regular Session).

Ragatz, R. 1979. *Housing Planning in Oregon.* Salem: Department of Land Conservation and Development.

Scott, Mel. 1969. *American City Planning Since 1890.* Berkeley: University of California Press.

So, Frank S., Irving Hand, and Bruce McDowell. 1986. *The Practice of State and Regional Planning.* Chicago: American Planning Association.

Sullivan, Edward J. 1990. "Affordable Housing and Growth Management in Oregon." Testimony to the Presidential Advisory Commission on Regulatory Barriers to Affordable Housing, San Francisco. August 2, 1990.

Whyte, William H, Jr. 1958. "Urban Sprawl." In *The Exploding Metropolis,* edited by the Editors of *Fortune.* New York: Doubleday.

CHAPTER 6
The Oregon Approach to Integrating Transportation and Land Use Planning

Sy Adler

This chapter analyzes the evolution of the Land Conservation and Development Commission's recently adopted statewide transportation planning rule. Administrative rule making, discussed in chapter 3 of this volume, was an effort by the state agency to clarify and define the methods local governments should use to achieve the objectives of Goal 12—the statewide transportation goal. This chapter provides a historical context for the emergence of rule making, discusses the structure and dynamics of the rule-making process, and concludes with an evaluation of the prospects for rule implementation.

The Origins of Rule Making

A 1987 study by the Metropolitan Service District (Metro) recommended that the Regional Transportation Plan be amended to include a new highway corridor as a preferred solution to transportation problems in the western portion of Washington County. Metro is the officially designated planning organization responsible for cooperative transportation decision making in the Portland region. Metro and Washington County staff, with the approval of the state Department of Land Conservation and Development (DLCD), agreed that the relationship of the proposed western bypass to statewide land use planning goals would be evaluated by the county during the course of project-level analysis. Crossing over, as it did, the urban growth boundary that Metro maintained for the region, the bypass raised several issues related to land use and transportation goals. If the proposed project failed to satisfy land use conditions, primarily the protection of agricultural and forest lands and other natural resources from urbanizing impacts, then it would not be built. Washington County incorporated the proposed highway corridor into its comprehensive plan, and Metro's Transportation Plan was similarly amended, both changes subject to the highway passing the relevant land use tests. Metro and Washington County decisions to proceed in this manner were challenged by Sensible Transportation Options for People (STOP), a citizen organization based in the bypass corridor, and by 1000 Friends of Oregon, the watchdog organization that has played a central role in the evolution of the state land use planning program.

STOP and 1000 Friends argued before the Land Use Board of Appeals (LUBA) that it was inappropriate for Metro and Washington County to defer an evaluation of whether or not the bypass was consistent with statewide land use and transportation goals to a project-level analysis. The petitioners claimed that such findings and arguments ought properly to have been made during the system-level planning process that had identified a highway as a preferred approach in the first place. Underneath their procedural critique lay a substantive concern: transport as well as land use alternatives that would either reduce or eliminate the need for a highway had not been thoroughly explored. LUBA basically agreed with STOP and 1000 Friends on the procedural issue, although the Oregon Court of Appeals later reversed LUBA's decision with regard to Metro (LUBA 1989a, 1989b; OCA 1990). However, while the legal issues were being sorted out, the Oregon Department of Transportation (ODOT), which would have to decide whether or not to commence its own study of the project, and Washington County sought guidance about the integration of land use and transportation planning from DLCD. The state land use agency decided to initiate a rule-making process aimed at clarifying relationships and preventing decision making delays regarding major facilities. The institutional product of this effort was the transportation planning rule adopted by the Oregon Land Conservation and Development Commission (LCDC) in 1991 to implement the statewide transportation goal. During the course of rule making, land use and transport planners had to confront longstanding issues of a general nature regarding the integration of these two aspects of urban growth, as well as a set of specifically Oregonian concerns.

A Historical Perspective on Projects and Plans

Transport projects and comprehensive land use plans have uneasily coexisted since the beginning of the modern urban planning movement. Leonardo Benevolo noted with regard to major mid-nineteenth-century European projects that "such was the urgency and complexity of technical demands that the overall planning potential of these new developments passed almost unnoticed and both legislation and practice acquired a specialized, departmental character, so that relations and connections between the various sectors were lost from sight. This was therefore unpromising terrain for the growth of town-planning legislation, and indeed the specialized legislation on railways and public works was later to prove one of its most powerful obstacles" (Benevolo 1971, 88). The "specialized, departmental character" of transport supply has, throughout the twentieth century, created profound uncertainties for U.S. urban planners who would—if they could—subordinate projects to the discipline of long-range land use plans. Projects designed by highway and transit engineers, however, often enjoyed powerful political support and had access to sources of implementation finance that were beyond land use planners' purview. As a result, projects typically ran far ahead of efforts to plan comprehensively. Given the relative weakness of metropolitan-wide planning, planners then sought to use transport projects indirectly to shape the pattern of urban growth.

Major facility-building programs in the 1950s and early 1960s crystallized the underlying tension between the state and regional agencies responsible for transport projects and those at the regional and local levels responsible for planning and regulating land use. Transport supply agencies faced unremitting demands for services and facilities from localities competing for investment, which intensified as population and economic activity within metropolitan areas dispersed. Roadway activists, looking back on the period of extensive urban highway construction during the 1920s, noted with chagrin that unregulated land development had transformed facilities intended as high-speed through routes into congested, property-serving local roads, rendering them functionally obsolete (U. S. House of Representatives 1944). State highway engineers, in particular, worried that the new freeways would meet the same fate; local land use regulatory regimes—succumbing to growth pressures—would undermine the integrity of the transport investment by permitting land uses that would overwhelm the facility (Sagamore Conference on Highways and Urban Development 1958).

Transit engineers were also concerned that the emerging pattern of suburban land development was inhospitable to the rail rapid transit systems they were designing. Low-density forms would likely complicate station accessibility for many potential patrons, and transit's capacity to shape future station area land use in patronage-enhancing ways would be compromised as well.

When the federal government massively accelerated freeway building in 1956, leaders of the urban planning profession saw an opportunity to structure metropolitan growth in accordance with the prevailing professional norm of functionally specialized, interdependent regions focused on a densely developed central business district. They were also acutely aware of the danger posed by the absence of planning capacity on the metropolitan periphery. They thought that in the bigger cities, where planners were established, relationships between land use and highway professionals would be cordial; joint efforts to integrate highways and land uses would be worked out. Outside the larger cities, though, until a planning capacity emerged, the highway engineers would be on their own. The planners worried that a narrow engineering approach would exacerbate tendencies to sprawl, thereby making urban form objectives more difficult to achieve (Howard 1957). Planners sought to enlighten the engineers regarding their profound responsibility for the future course of metropolitan growth, pointing out that transportation was much more than simply a function of land use. Transport investments also created land use patterns. When they made choices about the location and design of facilities, therefore, highway engineers were in fact functioning as urban planners (Webber 1959). State highway departments, for their part, strongly supported metropolitan-wide land use planning. In those areas lacking planning capacity, they advocated state legislative action to create it, although they also stressed their urgent mandate to build and the deeply troubling consequences of delay.

At meetings such as the Connecticut General Life Insurance Company's "The New Highways: Challenge to the Metropolitan Region" and the Sagamore Conference on Highways and Urban Development, planners and engineers worked toward a common understanding of their roles and relationships (Owen 1959, Sagamore 1958). In 1960 the federal Housing and Home Finance Agency and the Department of Commerce issued a joint policy and procedural statement encouraging cooperative and comprehensive approaches to metropolitan area development financed through federal highway funds and urban planning grants (Housing and Home Finance Agency 1960). In 1961 the American Institute of Planners and the Institute of Traffic Engineers followed with a joint policy statement regarding the appropriate division of professional labor in the urban transportation field, including both separate spheres of responsibility and shared tasks. Close cooperation in all phases was stressed in order to achieve a unified transportation program that would be fully integrated into a comprehensive land use plan (American Institute of Planners 1961).

While these professional and institutional accommodations were reached at the top, implementation on the ground remained problematic. This was due to intensifying competition between places within metropolitan areas, tightening resource constraints on transport suppliers, and an increasingly activist environmental movement that assumed the urban form banner that planners had held aloft in the early years following World War II. In Oregon, where an elaborate structure of state, metropolitan, and local land use planning has been in place since the middle 1970s, these dynamics still produced a great deal of uncertainty for both planners and engineers. Planning capacity had indeed emerged on the periphery of metropolitan areas, as all local governments were required by the state to adopt comprehensive land use plans. However, peripheral area plans aiming at promoting local growth proliferated. Coordinating these plans—and the transport projects they called for inspired by competition between places—in order to achieve metropolitan spatial form objectives remained problematic.

The Rule-Making Context

Rule making was framed by key contextual features, some of which were specific to the Portland area and others that were more general.[1] One was intensifying competition between downtown Portland and outlying business centers, especially those in the western part of Washington County. Business, political, and technical activists in these outlying areas sought transport projects that would facilitate autonomous, locally oriented growth, and either opposed or lacked interest in projects they saw as efforts to maintain the dominance of the Portland central business district.

Competition between places had greatly complicated the process of reaching consensus about regional transportation priorities. The Joint

Policy Advisory Committee on Transportation (JPACT)—an advisory body to Metro composed of local government representatives and transport supply agencies—had labored mightily during the latter 1970s and early 1980s to unite the metropolitan area behind downtown Portland-oriented westside light rail as the region's top priority project. In order to address mounting suburban demands, JPACT also strongly supported the bypass and other similar proposals for outlying areas. The STOP/1000 Friends challenge to the bypass threatened to disrupt the fragile consensus-based decision making regime that JPACT had constructed (Adler and Edner 1992).

During the 1980s, transport supply agencies, including ODOT and the Tri-County Metropolitan Transit District (Tri-Met) increasingly worried about their capacity to respond to future demands, given likely declines in federal subsidies and resistance to state and local tax increases. Moreover, transport suppliers and land use planners had to orient their efforts to achieving air quality objectives set out in federal clean air legislation, which called for actions to reduce travel demand. These objectives reflected increasing levels of environmental awareness and activism at all levels of government.

Metro projections that an additional 485,000 people would be moving into the Portland metropolitan area in the next twenty years, and that at the end of this twenty-year period—the year 2010—70 percent of all daily trips would be occurring within suburban areas, focused the issues. Much of this growth in population and travel was projected to take place in Washington County, which had been the locus of most of the region's increase in the past decade. Concerns about this coming boom galvanized a host of state, regional, and private agencies whose agendas would be shaped by demands associated with it. There were similar, though less intense, growth pressures in other urban areas in the state as well. The capacity to manage growth pressures and maintain the integrity of the growth boundary approach to achieving urban form objectives was facing the first serious challenge since the statewide land use planning structure was put in place. The impacts of the prevailing pattern of suburban development in the Portland area and elsewhere on the demand for energy and the achievement of air and water quality and transport goals were highlighted as well.

The Metro projections were often linked in the media and in public pronouncements with negative references to worsening traffic, air quality, and sprawl-related problems in Seattle and California cities. A recent successful campaign to secure local funding for westside light rail, for example, had hammered on this comparison. While the specter of Los Angeles was a prominent theme in public discourse, though, 1000 Friends, STOP, the local chapter of the Sierra Club, and other environmental organizations feared that the pattern of permitted suburban growth—which they saw giving rise to the demand for the bypass—was leading the region in precisely that direction.

Taken together these contextual elements called into question the formal and informal arrangements through which participants in land use and transportation planning had related to each other. Resolving the ambiguity about when in the planning process statewide land use and transportation goals ought to be addressed initially framed rule making. More general concerns about the relationship of projects to plans, intergovernmental relations, and the achievement of substantive urban form and transport objectives would become critical agenda items as well.

The Structure of Rule Making

The structure of rule making featured an alliance at the top between the state-level governmental agencies, DLCD and ODOT, a divided local government sector below, and critical support for the state-level alliance from activist citizen environmental organizations. During the early stages of rule making, the state agencies were primarily interested in answering procedural questions and in clarifying intergovernmental relationships in the transportation planning process. However, DLCD and ODOT were also increasingly worried about the capacity of local government agencies to play their assigned role in achieving state agency objectives regarding urban growth management and transport system planning. They were interested in limiting the discretion available to local actors in order to focus local attention on higher level goals. State-level concerns about local capacity were shared by 1000 Friends, which had devoted much of its energy over the years to a litigation-based approach to monitoring governmental plan making and implementation.

During the 1980s ODOT was increasingly occupied with the land use dimension of its highway facilities; the engineers' long-standing interest in protecting the integrity of the highway investment was once again on the agenda. Personnel changes within the agency as well as new appointments to the Oregon Transportation Commission—the policy-making body for ODOT—led to a renewed interest in land use. In addition, ODOT was moving into new areas: carrying out a legislative mandate to prepare a statewide multimodal transportation plan; allocating funds to designated interurban access Oregon highways; and programming the modernization of its roads. In these various ways, ODOT was reasserting a leadership role in transport planning. This thrust followed a period during which the agency had tried largely to respond to local initiatives, a mode of action which itself had been a response to the political opposition generated by the postwar wave of bold freeway building. Moreover, as they contemplated their new tasks, ODOT personnel had their own version of the Los Angeles specter: highways whose functional integrity and level of service had been fatally compromised by incompatible land use activities along the route, as in the case of Highway 97 through Bend, a growing resort town in the central part of the state, where a road intended as a bypass now served an intensely developed commercial strip. ODOT was also troubled by local conflicts that stymied or substantially delayed its efforts to plan and construct highways of statewide significance; these critical routes were sometimes held hostage to land use-related local political disputes.

ODOT, therefore, was interested in a rule that would enshrine the dominant role of its statewide transportation systems plan. Local plans and plan amendments should incorporate the routes that ODOT deemed to be of critical statewide significance, facilitate their construction, and maintain their functional integrity. ODOT realized that acting independently it was in a relatively weak position to manage access to its roads; the rule should require local governments to assist, as well as constrain local ability to approve land use changes that would degrade the level of service. At the same time that the agency was participating in rule making, ODOT was also rewriting its state agency coordination plan, which would address the way it related to local governments regarding land use issues. ODOT would attempt to secure its position there as well.

DLCD supported ODOT's quest to establish its dominant position in transportation planning, and sought to elaborate its own leadership role in addressing related land use and environmental issues. At the same time that the state land use agency was engaging in rule making, it was also pursuing a growth management study, and participating with others in a state agency growth council concerned with the Portland metropolitan area. The department worried that local growth management regimes were not working as they should, and that growth boundaries were not sufficient to achieve urban form objectives. However, DLCD had to confront the existence of local comprehensive plans that the agency had itself acknowledged to be in conformance with statewide land use goals. Ironically, the Washington County transportation plan—the target of the 1000 Friends legal action—was regarded by the agency as the best such local effort in the state; it was the exemplar that was held up for others to emulate.

Though LCDC's rule would set the framework within which local governments would integrate transportation and land use planning, some local governments, especially Washington County, vehemently argued that there was a legitimate hierarchical relationship that the rule ought to respect: transport plans should be subordinate to and implement comprehensive land use plans. The local governments sought to protect the integrity of acknowledged land use plans that had been so politically complicated to produce. In addition, some downstate local governments feared the imposition on all of them of yet another set of detailed and costly requirements that were really addressed only to problems in the Portland area and a few other parts of the state. Finally, local governments generally applauded ODOT's new initiatives, particularly its multimodal planning efforts. However, they also believed that ODOT ought to pay much more attention to intraurban transport needs than it had in the past.

1000 Friends, STOP, and their environmental allies had two related substantive interests that they constantly pressed forward during rule making. One was that land use and transportation plans conform to all dimensions of the statewide transportation goal including, in particular, the directive to "avoid principal reliance upon any one mode of transportation." Linked to this was their concern to create an urban form that would facilitate the achievement of the modal objective. The environmental activists sought a compact, densely developed urban region—the land use planners' traditional desire—in order that other modes of transport, including transit, walking, and bicycling, might diminish the dominance of the automobile-highway system.

The environmentalists perceived that land use planners in outlying areas had lost sight of their historic urban form objective; they had become caught up in the competition to attract and hold investment in their local jurisdictions. Therefore, the activists felt justified in insisting that local governments review their land use and transportation plans—plans which had been acknowledged by the state—with both state and local eyes focused on whether or not the plan in question produced a spatial pattern of growth that would avoid principal reliance on the automobile.

1000 Friends clearly challenged the priority accorded to acknowledged land use plans by local government. Regarding Washington County's acknowledged comprehensive plan and the relation of the western bypass to it, the Friends believed that "the land use designations in the county's plan are no more sacred than the transportation facilities. There is no legal or policy reason why land use designations must remain fixed while a multimillion dollar highway is retrofitted in to the plan." Not only did the environmentalists take a much more flexible approach to acknowledged plans, they also sought to transcend the place-based competition for investment that was such a key contextual feature of the transportation planning process. In order to reduce dependence on the auto in Washington County 1000 Friends was willing to ask: "But what if there was a change in the amount and type of development which occurred among the different planning zones? Perhaps if some of the residential development were concentrated further east and if some of the dispersed commercial development were concentrated in Beaverton [in the eastern part of the county], the result would be a somewhat different pattern of trips. . ." The substantive transport and urban form objectives were supported by the state energy and environmental quality agencies, as well as by City of Portland planners.

Structural elements and contextual features combined to orient rule making toward a fundamental rethinking of transport and land use plans, and their integration. Intervention-minded state agencies, conscious of increasingly demanding environmental mandates and constraints on transport facility supply, looked to reshape the field of local action. The longevity of the Oregon system of planning also imparted a special character to rule making. State and local agencies and citizen organizations had been working with each other on transport and land use issues for several years. Many of the professionals involved in rule making had been associated with each other in a variety of forums, and had established working relationships. The state system provided participants a shared vocabulary and, in many cases, a set of shared objectives, although differences rooted in structural positions remained. While in other parts of the country public and private actors were motivated by environmental and transport supply crises to address the issues that would be covered during rule making, state and local professionals in Oregon saw themselves as anticipating problems and attempting to prevent crisis conditions from developing, and the state's institutionalized, adaptable planning system made this possible. Rule making would clearly be an elaboration of existing sets of procedural requirements and substantive objectives, even as radical departures from extant practices were contemplated.

The Dynamics of Rule Making

The dynamics of rule making featured a steadily broadening field of vision from the procedural questions initially of concern to the DLCD-ODOT alliance and local governments to the substantive issues raised by environmental activists. DLCD orchestrated rule making. ODOT signaled the seriousness of its commitment by hiring a land use attorney—formerly associated with 1000 Friends—to facilitate the process. DLCD and ODOT together produced concept papers and drafts of the rule. These papers and drafts were circulated to local government transportation planners, transport supply and environmental organizations, and other interested individuals and groups. DLCD received written comments from these participants, and these same parties had the opportunity to offer verbal testimony when the rule was discussed at LCDC meetings. In addition to these possibilities for written and verbal input, DLCD sponsored a set of workshops to which the parties were invited. An LCDC member who was a land use planner facilitated these workshops, and represented the commission throughout rule making. DLCD and ODOT were not aiming to build a solid consensus among all the participants, and some local governments voiced a concern that the process was not as participatory as it ought to have been. DLCD and ODOT staff did, however, continually redraft the rule in relation to the public comments. From the middle of 1989, when the process began, to the early part of 1991, when a final draft rule emerged, the participants moved as a group, despite continuing doubts and disagreements, to support a greatly expanded regulatory regime aimed at achieving substantive urban form and transport objectives.

At the outset, in a June 1989 memo on the subject, DLCD referred to their effort as aiming at the creation of a "highway" planning rule that would be adopted by January 1990. The key concern was siting highway improvement projects outside urban growth boundaries; a rule would guide decision making on projects being planned in relationship to the statewide land use goals for land use planning, public facilities, and urbanization. In the relationship between land use and transportation plans, DLCD gave priority to acknowledged comprehensive plans setting out land uses; the role of transport was seen as supporting the objectives expressed therein.

Shortly thereafter, though, DLCD was informed by ODOT and Washington County that swift action on a rule was no longer necessary, since ODOT had decided to embark on a multiyear study of traffic issues in western Washington County. ODOT encouraged DLCD to take up the coordination of transport and land use planning in a more comprehensive manner, as did STOP and 1000 Friends. DLCD agreed to do so, and began to look toward rule making under the statewide transportation goal.

In response to the procedural question of when to address consistency with statewide land use goals in the planning process, rule making converged on a distinction between system- and project-level planning, with goal analyses required at the level of transport system planning, except under certain circumstances. Metro, Washington County, and DLCD had originally agreed to defer goal analyses to the project planning stage of the western bypass; Metro argued during rule making that it was important to build in such flexibility to the planning process because critical sorts of information might not surface until a more specific stage. STOP and 1000 Friends, joined by some local governments and state agencies, believed that it would be too difficult to do a thorough goal evaluation at this point. Once identified, they argued, projects acquired momentum toward construction that was very difficult to counter. The environmental activists wanted statewide goals to shape regional and local commitments to specific land use patterns and transport projects as much as possible.

DLCD and ODOT now wanted statewide goals addressed during system planning. However, DLCD proposed a compromise position. Recognizing Metro's concern, goal analyses might be deferred to a refinement plan if a local or regional planning agency demonstrated a lack of information necessary to make a final determination, showed that deferral did not affect the integrity of the rest of the system plan, and that the refinement plan would be completed in a timely manner. The activists' concern with project momentum was addressed with a set of substantive requirements for the preparation of system plans, discussed below.

The state-level position on related procedural questions regarding the location of transport facilities on rural lands reflected the desire of DLCD, ODOT, and 1000 Friends to limit local government discretion. The alliance wrote into the rule a long list of facilities and improvements that would be defined as consistent with statewide agricultural, forest land, public facility, and urbanization goals, and therefore permitted on rural lands without condition, and a set of improvements that would be permitted if conditions specified in the rule were met. Facilities and improvements that did not meet these conditions would require an exception to the statewide goals. The alliance also wrote into the rule a lengthy set of requirements for justifying an exception. DLCD, ODOT, and environmental activists were clearly concerned about the capacity of local governments to resist the land development pressures that often accompany transport investments. The extensive level of detail, reducing the amount of discretion available to local planners, aimed at protecting ODOT's transport investments as well as at achieving the urban form objectives of DLCD and the environmentalists. Requiring local planners to seek exceptions—which the alliance chose in the face of opposition to this approach from some local governments and development groups—would open up local decision processes to monitoring by state and environmental organizations, thereby also limiting discretion.

The alliance did divide on one aspect of this issue, though. Bypasses are permitted to cross over urban growth boundaries onto rural land if trips within the urban growth boundary would account for fewer than one-third of average daily traffic using the facility. 1000 Friends and STOP thought the one-third figure was too high, and suggested that 10 percent would be appropriate for a road that would truly be a bypass. DLCD and ODOT refused to make the change.

The centrality of ODOT's statewide transportation system plan clearly emerged during rule making, and ODOT's concerns regarding the protection of its investments were addressed. However, an element of ambiguity remains regarding the relationship of state transportation projects and comprehensive land use plans. Regional and local transport system plans must be consistent with ODOT's statewide plan. In this sense, the rule seeks clearly to establish a hierarchically organized set of intergovernmental relations. ODOT's transportation projects will also have to be compatible, though, with acknowledged comprehensive plans. LCDC here sought to maintain the integral role of the comprehensive land use plan in the planning process. ODOT's state agency coordination plan is referenced in the rule as the framework within which any intergovernmental conflicts are to be resolved; the procedures for conflict resolution are very general in nature. As in the case of refinement plans, the extent of the ambiguity in the relationship between state and local levels is circumscribed by the requirement that both state and local plans be consistent with substantive urban form and transport goals.

Rule making clearly established the importance of maintaining the functional integrity of transport investments. Local governments are required to adopt regulations to control access, protect future operations, minimize the impacts of development proposals, and give ODOT and other transport supply organizations timely notice of any possible land use activities of relevance to them. Local planners must also insure that amendments to adopted plans and regulations are consistent with service levels, functions, and capacities of facilities that are identified in transport system plans.

In addition to ODOT's central role, the rule requires that mass transit, transportation, airport, and port districts participate in system planning processes, and prepare their own plans consistent with system products. Local planners must implement land use and design regulations to protect airport operations; provide bikeways, bicycle parking facilities, and pedestrian ways in various development contexts; and support transit service by designating land for transit oriented development along routes, and mandating major developers to provide either a transit stop or a connection to a stop when the service provider requires it.

The substantive goals and requirements in the rule clearly reflect the weighty presence of 1000 Friends and their environmental activist partners; as one participant put it, they "pushed the envelope." From the very beginning of the rule-making process 1000 Friends argued that both land use and transportation plans ought to be consistent with the Goal 12 requirement to reduce principal reliance on the automobile. The activists insisted at the outset that acknowledged comprehensive plans be reexamined—and changed—if their implementation necessitated continued dependence on the auto-highway system. A mandate to reconsider acknowledged plans was one of the most controversial aspects of rule making, and the last major element of the rule to emerge.

The Friends suggested early on that the rule require transport plans to set targets for mode shares, and then identify facilities and land use patterns, and demand management policies to achieve them. While rule making did not converge on that particular target, the idea of a measurable objective resonated with the LCDC member who was working on the rule. Recalling the commission's experience with the implementation of its statewide housing goal, the commission member believed that the absence of numerical targets—which had been the case in housing—had hindered goal achievement. The commissioner urged the participants to define an approach to reducing principal reliance on the auto that could be quantified and incorporated in the rule.

DLCD and environmental activists sought to use numerical targets to further limit the discretion available to local government planners; this effort was, in turn, resisted by those responsible for implementation. A politics of numbers evolved, with an evident tendency to compromise in the name of flexibility. Competition between central business districts and suburban business centers was clearly one of the issues.

Early drafts of the rule required local and regional system plans to meet three standards to achieve reduced reliance on the auto in metropolitan planning organization (MPO) areas: doubling the share of non-auto modes; an auto occupancy rate of 1.3 persons during commute hours; and reduced vehicle miles traveled (VMT) per capita. At this point the rule did not specify a target for VMT reduction. DLCD expected "considerable comment on these proposed standards, both that they are too strict and not strict enough." The agency was right. Several local planners questioned the feasibility of attaining these standards, while 1000 Friends suggested that the rule also require a doubling of the modal share of non-auto work-related trips, and a higher auto-occupancy target.

Rule drafts also required local governments to design and implement a parking plan that specified limits on the number of parking spaces. Ratios relating parking spaces to the number of employees and the amount of retail floor space within central business districts and transit oriented development districts, and in other employment and retail areas, would be spelled out. Once again, some local planners questioned whether or not these ratios made economic sense, especially in suburban areas, while other parties worried that the limits were too generous.

The City of Portland's Transportation Office, for example, argued that "the proposed ratios of parking space to floor area or employees is so generous that it is unlikely to make any improvement in the situation. Further examination of existing parking ratios in suburban areas is needed to determine a reasonable requirement that would actually reduce the amount of parking that is currently constructed.... The proposed ratio would double the amount of parking allowed in the Portland CBD [central business district]." 1000 Friends strongly supported parking limitations, but also felt the ratios in the draft rule allowed too much. With their urban form and transport goals clearly in mind—and lacking interest in the dynamics of spatial competition—they noted that "in order to encourage development in those areas best served by transit (i.e., CBDs and TODs [transit oriented districts]), ratios outside CBDs and TODs must not be substantially higher than those inside the CBDs and TODs. Otherwise, there would be a significant advantage for new development to be located outside of CBDs and TODs."

DLCD responded to these conflicting pressures by proposing to specify in the rule the single standard of a 20 percent reduction in VMT per capita, to be achieved over the course of a twenty-year planning period. Local planners would then decide the particular combination of elements, including increased non-auto mode share, ride sharing, demand management, and parking that made the most sense in particular circumstances. DLCD also proposed to shift from the detailed ratio approach to mandating a 20 percent overall reduction in the number of parking spaces per capita in the whole MPO area. 1000 Friends responded that this would be more difficult for local governments to implement than the ratio idea since they would have serious trouble measuring spaces per capita and enforcing compliance. DLCD refused, however, to alter its revised stance.

In its final recommendations to LCDC, the department was still more forthcoming toward local planners. In order to moderate resistance and increase the likelihood of innovative responses, DLCD proposed that the VMT reduction goal be stretched out to thirty years, and the parking reduction target scaled back to 10 percent. Planners are required, though, to specify interim benchmarks for non-auto mode share, auto occupancy, and demand management to measure progress toward VMT reduction, and to evaluate at five-year intervals. The rule requires that system plans be amended in the absence of progress in this area.

Three local planning directors in the Eugene area forecast a dark side to the numerical target approach: "Partially because of the practical difficulties [in measuring VMT and population within MPO areas] the unintended effect of this proposed standard will be the 'numbers game.' Local agencies may be encouraged to adjust their models to achieve results required by the rule, rather than produce their best estimates of anticipated demands on local infrastructure" (Burns, Childs and Daluddung 1991). Acknowledging the difficulties raised by participants, particularly the lack of evidence about the feasibility of changing travel behavior by the specified amount, and about measuring changes, LCDC committed itself to revisit the target every five years to check on its continuing validity.

One of the primary objectives of environmental activists was required examination of alternative land use plans as a strategy to reduce principal reliance on the automobile. The rule incorporated a mandate for local governments in Metro's planning area to do so; it is optional for others. The strategy requires consideration of: increasing densities and specifying minimum residential densities near transit lines and near large employment and retail shopping areas; increasing densities in new office and retail projects; mixing residential and neighborhood commercial land uses; designating land uses to achieve a closer balance between jobs and housing; and setting maximum parking limits at office and institutional developments which shrink the supply of parking available there. Reducing principal reliance on the auto is one of the criteria to be used to evaluate the alternatives considered, along with various other environmental, energy, and transport system connectivity objectives, and a requirement to serve land uses identified in acknowledged comprehensive plans.

DLCD and ODOT proceeded very cautiously on this issue. A substantial concern for ODOT was that a required reexamination of alternative land use plans might result in protracted political stalemate, significantly delaying construction of projects ODOT thought important. DLCD was obviously aware of the amount of effort that went into producing—and the political fragility of—those plans that had been acknowledged. In addition, local planners articulated profound philosophical as well as practical concerns about the relationship of transport and land use planning to DLCD.

At the outset of rule making, in response to an early concept paper, Metro's legal counsel charged that "If . . . [an] acknowledged land use plan is required to be reconsidered . . . this is a radical concept for rule-making. The existing land use structure labored for years to establish acknowledged comprehensive plans. Transportation planning is to support such plans." Washington County's planning manager agreed that a requirement to redo land use plans was an unnecessarily radical step, "equivalent to assigning to the scrap heap the fifteen years of local planning and the state acknowledgment review process. . . . The DLCD rule concept clearly fails to grasp that . . . land use plans come first. Transportation plans are in second priority to land use plans; transportation plans require the preexistence of land use plans and are explicitly designed to support land use plans." The Association of Oregon County Planning Directors added that "the proposed rule should not give transportation planning priority or equal status with the overall comprehensive planning process."

DLCD disagreed in principle with these arguments. During the initial stage of rule making the department argued that "few acknowledged plans were based on a long-term assessment of the transportation system needed to support the proposed land use pattern. . . . Good comprehensive planning should allow for reconsideration of the proposed land use pattern if it cannot be supported by the transportation system." Draft rules, however, made reconsideration an option rather than a requirement. DLCD agreed with much of what 1000 Friends was saying about the importance of changing land use patterns if the goal of reducing principal reliance on the auto was to be achieved. The department believed, though, that rule mandates to plan for better bicycle, pedestrian, and transit connections, and to zone land for uses supportive of transit and developments oriented towards transit went a long way toward a full-scale reexamination. In addition, land use patterns would be reconsidered if interim benchmarks for reduced auto reliance were not met.

DLCD also believed that "a state requirement to reconsider land use patterns should reflect all of the relevant policy objectives, not just transportation objectives. . . . [A] requirement that local governments reconsider land use patterns should respect the degree of effort and difficulty involved. Serious reconsideration of land use patterns by local governments will be a major planning task. If it is to be done, it should be a thoughtful and comprehensive effort done on its own right, not simply as an element of a transportation plan." The department noted its own study of growth management then taking place as an argument against requiring reexamination at that time.

Broader political currents began swirling around this aspect of the rule. Twenty-eight members of the Oregon legislature, including much of the Portland area delegation in both the House and the Senate, wrote the LCDC chair to "adamantly urge the Commission to amend . . . [the rule] . . . to require, not merely allow, the consideration of land use alternatives." The chair of the Oregon Environmental Quality Commission followed shortly thereafter: "I write to ask that the permissive language . . . of the Plan [regarding the reconsideration of land uses] be made mandatory. You have heard from other Oregon policy-makers to the same effect." During the rule-making period, Metro had embarked on a process to formulate regional urban growth goals and objectives, which involved an evaluation of land use patterns. The agency shifted its position to support reconsideration as part of the rule. Tri-Met and the City of Portland also agreed. Within the region, Washington County was increasingly isolated on this issue.

DLCD remained reluctant, though, actively to embrace required re-examination. When the final draft was presented to LCDC for adoption, the department did not propose that the rule include the requirement. However, DLCD pointed out that the commission might wish to require reconsideration in the Portland metropolitan area. The department noted that influential support for a mandate had crystallized, and reminded the commission that there was a precedent for a Portland area focus in the Metropolitan Housing Rule. LCDC decided to incorporate such a focus in the Transportation Planning Rule as well. In a significant break with the past, though, the rule permits cities with fewer than 2,500 residents and counties with fewer than 25,000 to apply for an exemption from the requirement to prepare a local transportation system plan. DLCD was here clearly indicating that it had learned an important lesson: the entire statewide land use planning system need not gear up to respond to problems that were acutely felt only in the major population centers.

Retrospect and Prospect

The rule orients transportation and land use planning in metropolitan areas toward reducing principal reliance on the automobile. DLCD noted that this particular transportation goal requirement was the only substantive, mode-specific element; otherwise, the goal was mode-neutral. Changing land use plans and implementing ordinances as a way of changing travel behavior, however, was clearly controversial; there was and remains much skepticism about both the theory and practice. The City of Salem assistant planning administrator argued at the outset of rule making that ". . . use of land use as the backbone of a process to create a shift in modes appears simplistic. Such changes are due to more than a shift in land use" (Budke 1990). The City of Gresham's community development director pointed out that the rule itself would have little to do with the financial capacity to implement transport supply alternatives in keeping with the rule's mandates. "Rewriting local land use rules will make little difference if statewide transportation revenue sources and programs continue to place the overwhelming emphasis on auto transportation" (Andersen 1991). The state's gas tax remains, for example, constitutionally limited to highway-related expenditures. The 1991 federal surface transportation act, though, increases the capacity of state and regional transport planners to shift federal gas tax revenues between modes. A transportation planner in the Eugene area added that "increasing residential densities is no guarantee of shifts of automobile trips to other modes. Without significant pricing and public policy intervention, Americans have shown an amazing adaptive capacity to remain in their automobiles" (Gordon 1991). The Eugene planner also pointed out that the strategy of increasing densities to achieve urban form and transport objectives entailed a number of costs that the rule did not address: the likely resulting increase in land prices would make other planning goals—affordable housing and open space preservation, for example—more difficult to achieve. This planner clearly articulated the sense of many colleagues at the local level who felt that the land use planning process had been hijacked to meet narrow transport objectives. "If densities and land use patterns are to be increased to achieve transportation goals, then the trade-offs and negative impacts must be taken into account. We should take care not to approach such a complex issue within a 'one goal' context. We should also take care not to allow any particular goal, taken in isolation from other goals, to dictate our comprehensive land use direction. Only through the comprehensive land use context can all the positive and negative effects of major State policy change be evaluated within different communities" (Gordon 1991).

There is support in the academic literature for a skeptical view of the rule's prospects. Robert Cervero has called upon planners to alter land use patterns in order to deal with traffic congestion, but he clearly points

Integrating Transportation and Land Use Planning 141

out that "the best solution would be to price low density living correctly through higher property taxes, fuel taxes, and congestion fees." Cervero goes on to argue, though, that "because of equity concerns and political inertia, congestion charges and 'sprawl' taxes have yet to materialize in the United States. This, then, leaves land use practices as more or less a second-best solution to the problem" (Cervero 1991, p. 120). He calls for planners to "seize the opportunity to shape land development while powerful macro-changes continue to unfold" (Cervero 1989, p. 148).

Cervero notes that in those few cases where efforts are being made to integrate transport and land use planning—New Jersey and Florida, for example—state-level leadership and mandates were critical to force local governments to act. "By linking state aid and infrastructure funds to coordinated planning and by enforcing federal laws regarding environmental protection and housing discrimination, these and other states are beginning to force a structure of coordinated planning upon localities, regional agencies, and their own state bureaus" (Cervero 1991, p. 126). State agencies have clearly taken the lead in Oregon, and have imposed a very detailed structure on local and regional planning practice. Responsiveness to federal environmental mandates is clearly in evidence, and there is a high level of awareness regarding the distribution of affordable housing. Rule makers consciously chose, however, not to incorporate a strict concurrency requirement linking infrastructure finance and land development, as was done, for example, in Florida. The rule requires local planners to prepare transportation financing programs. It mandates that these programs provide for phased infrastructure projects that will encourage infill and redevelopment rather than cause premature urbanization. However, timing and finance provisions in these programs are not to be considered land use decisions, and cannot be the basis of an appeal. Rule makers were concerned that a strict concurrency requirement would be too rigid an approach, and might undermine efforts to achieve urban form goals if new development sought out peripheral areas where infrastructure capacity was readily available. The rule maintains the separation between planning and implementation finance that has characterized the Oregon planning system since its inception.

Cervero suggests four promising land use solutions to transportation problems: 1) increased density; 2) mixed land use; 3) balancing jobs and housing; and 4) pedestrian-oriented site designs (Cervero 1991). The rule enthusiastically embraces all of these. Elizabeth Deakin notes, though, that "[T]here remains considerable disagreement about whether . . . revis[ing] general plans, subdivision regulations and zoning to provide for development patterns and levels that help reduce overall automobile use" will be effective (Deakin 1989, p. 85). Deakin points to the very long-term nature of the impact of such changes, as well as to the possibility that trips in the range of 3 to 10 miles—the sorts of trips that would be likely outcomes of more balanced jobs and housing location patterns—would be too short for ride-sharing schemes to be attractive, and too long for walking. The rule requires MPOs to complete regional transportation systems plans within four years, and local governments within MPO areas to adopt system plans and implementing ordinances by one year later. However, in order to shorten the implementation timeframe, cities and counties of 25,000 or more population must put in place regulations supporting developments oriented towards bicycle, pedestrian, and transit use within two years following the adoption of the rule.

In a cautionary survey of southern California land use-based initiatives to deal with congestion, Martin Wachs emphasizes that research about these sorts of efforts is lagging very far behind practice; there is much uncertainty about the capacity of available techniques to monitor and evaluate the outcomes of these programs as well (Wachs 1989/90). Local planners and state officials were acutely aware of these difficulties during rule making, as reflected in the provisions for interim benchmarks and LCDC's commitment to evaluate the VMT reduction targets after five years. Cervero adds that ". . . municipalities are continually vying for attractive land developments. . . . Clearly, any successful joint land use and transportation planning effort will hinge on finding ways of moderating the competitive and parochial instincts of local governments" (Cervero 1991, pp. 126-127). While moderation is implicit in the 1000 Friends approach to integration, the rule itself is silent on this critical point.

The literature is clear regarding the viability of one form of intervention, though: parking management works. Surveying the demand management experience, Erik Ferguson concludes that "parking management, particularly parking pricing, has been found to have the largest and most consistent impacts among transportation demand management elements" (Ferguson 1990, p. 452). Pointing to the same finding, Wachs adds that "the vast majority of [transportation demand management programs] had little or no effect on commuting behavior, and virtually all of those that were ineffectual left subsidized employee parking benefits intact" (Wachs 1991, p. 336). The rule mandates the implementation of a parking space reduction plan, and the inclusion of minimum and maximum parking requirements in land use regulations. The rule does not specifically address actions to alter parking prices. A trip reduction ordinance—which typically requires employers to change employee travel behavior, often through parking management programs—is offered as an example of demand management; the rule leaves it as a local option.

DLCD and ODOT, urged on by environmental activists, have woven a tightly fitting garment for local and regional planners to wear. The Oregon approach to land use and transportation problems is a planning approach; the rule constitutes a major elaboration of this planning system. LCDC has the power to structure the plans that local and regional planners must produce. It has used its power in this case, as it has in others, to attempt to strengthen the capacity of local governments to achieve urban form and transport goals in the face of growth pressures threatening to undermine the possibility of success. ODOT saw the rule as a way similarly to shore up the local capacity to protect the integrity of their investments. 1000 Friends has long deployed its legal resources as a counterweight to these pressures when it perceived a local failure to defend and advance the statewide interest.

The rule reflects the perspective of these statewide actors. Given their shared concern with local susceptibility to capture by developers and political officials pursuing short-term gains, they wrote a detailed rule that aims to reduce the discretion available to local planners in the content of the plans they prepare and the ordinances they must draw up. Uncertainties regarding implementing persist, though. The manner in which the state transportation system plan will mesh with its regional and local counterparts remains to be worked out. The rule was written by professional land use and transportation planners. It remains to be seen what land developers—responding to market forces—think about, for example, the increased densities, mixed uses, and parking reduction mandates. The reaction of neighborhood associations to increased densities and the other regulatory and design changes also remains to be registered.

DLCD believes the standards in the rule are attainable. The agency is counting on the accumulated good will of the planning community—and the general commitment to planning throughout the state—to attain them. The department is also well aware of the likely limits of a land use approach to changing travel behavior. The rule is an effort to stimulate a wholesale state and local policy shift to prevent the environmental and transport crises that beset other parts of the nation. As DLCD put it in its recommendation to adopt the rule: "Continued reliance on the automobile . . . means reduced mobility, traffic congestion and reduced air quality and ultimately even more expensive and difficult measures to deal with these problems. Land use planning has an important leadership role in addressing this problem. Comprehensive plans express a vision for development over the next twenty years. If land use plans are based on continued auto reliance, implementation of other measures to achieve reduced reliance will be made more difficult."

The Oregon transportation planning rule is the first concrete manifestation of the understanding in principle worked out between leading land use and transportation planners a generation ago. Oregon planners now begin translating central tenets of planning theory into practice. A thorough evaluation of this effort is entirely appropriate.

Notes

1. The discussion of rule making is based on interviews with participants and on documents supplied by them.

References

Adler, Sy, and Sheldon Edner. 1992. "Challenges Confronting Metropolitan Portland's Transportation Decision System." *Transportation Research Record No. 1364*. Washington: National Academy Press.

American Institute of Planners and Institute of Traffic Engineers. 1961. "Professional Responsibility of City Planners and Traffic Engineers in Urban Transportation: A Joint Policy Statement." *Journal of the American Institute of Planning* 27, 1: 70-73.

Andersen, John. 1991. Letter to Susan Brody, DLCD. January 14. On file with the author.

Benevolo, Leonardo. 1971. *The Origins of Modern Town Planning*. Cambridge: MIT Press.

Budke, Roger. 1990. Letter to Bob Cortright, DLCD. July 12. On file with the author.

Burns, Roy, Jan Childs, and Susan Daluddung. 1991. Letter to Robert Courtright, LCDC. April 4. On file with the author.

Cervero, Robert. 1989. "Jobs-Housing Balancing and Regional Mobility." *Journal of the American Planning Association* 55, 2: 136-150.

———. 1991. "Congestion Relief: The Land Use Alternative." *Journal of Planning Education and Research* 10, 2: 119-129.

Deakin, Elizabeth. 1989. "Land Use and Transportation Planning in Response to Congestion Problems: A Review and Critique." *Transportation Research Record*. No. 1237: 77-86.

Ferguson, Erik. 1990. "Transportation Demand Management: Planning, Development, and Implementation." *Journal of the American Planning Association* 56, 4: 442-456.

Gordon, Steve. 1991. Memo to Planning Directors. April 3. Lane Council of Governments. On file with the author.

Housing and Home Finance Agency - U.S. Department of Commerce. 1960. "Joint Policy and Procedural Statements on Improved Coordination of Highway and General Urban Planning." In *Readings in Urban Transportation*, edited by George Smerk. Bloomington: Indiana University Press. 1968. 251-256.

Howard, John. 1957. "Impact of the Federal Highway Program." *Planning 1957*. Chicago: ASPO. 35-41.

Land Use Board of Appeals (LUBA). 1989a. 17 Or LUBA 671 (1989). 671-691.

———. 1989b. 18 Or LUBA 221 (1989). 221-252.

Oregon Court of Appeals (OCA). 1990. 100 Or App 564 (1990). 564-570.

Owen, Wilfred. 1959. *Cities in the Motor Age*. New York: Viking Press.

Sagamore Conference on Highways and Urban Development. 1958. *Guidelines for Action*. Syracuse University.

United States House of Representatives. 1944. *Interregional Highways*. Document #379. Washington: Government Printing Office.

Wachs, Martin. 1989/90. "Regulating Traffic by Controlling Land Use: The Southern California Experience." *Transportation* 16, 3: 241-256.

———. 1991. "Policy Implications of Recent Behavioral Research in Transportation Demand Management." *Journal of Planning Literature* 5, 4: 333-341.

Webber, Melvin. 1959. "The Engineer's Responsibility for the Form of Cities." *Traffic Engineering* 30, 1: 11-14, 39.

CHAPTER 7
Siting Regional Public Facilities

Mitch Rohse & Peter Watt

> *"The test of a first-rate intelligence is the ability to hold two opposed ideas in the mind at the same time, and still retain the ability to function."*
>
> F. Scott Fitzgerald, *The Crack-Up*

Even after two decades, Oregon's statewide planning program remains controversial. Much of the controversy stems from the basic design of the program, which rests on two opposing ideas.

The first idea is that because the use of land often affects vital state or regional interests, the state should assert a strong role in land use planning. Many legislators supported Senate Bill 100 in 1973, for example, because they saw rapid development of farm land as a threat to the state's agricultural economy.

The second idea is that land use planning is best done by local governments—by the officials closest to the land and to the citizens who use it. This (and political concerns about state versus local control) led the designers of Oregon's planning program to specify that city and county comprehensive plans should be the controlling documents for all land use decisions.

For the most part, the tension between those two opposing ideas has been a productive force. It has brought state and local officials together in a partnership for planning. That partnership continually strives to find a balance between the uncoordinated, parochial planning found in some states and the cumbersome, top-heavy systems found in others.

Preservation of farm land is one example of Oregon's successful balancing of state and local interests: the state defines agricultural lands and sets the standards for preserving them; counties plan and zone the agricultural lands, and administer all permits for development and division of such land.

Such a balance has not been achieved, however, in the siting of regional public facilities such as prisons.[1] Instead, two separate and sometimes conflicting siting processes have evolved. One, growing out of Senate Bill 100, emphasizes the state-approved local plan as the controlling document for all land use decisions. The other, based on a variety of "supersiting" laws, exempts regional facilities from some or all local planning and zoning requirements. The two processes are contrasted below.

Siting Regional Facilities through the Acknowledged Local Plan

Senate Bill 100 established a system that gives cities and counties the authority to review and decide on applications for land use permits. Under that system, conditional use permits, variances, rezonings, subdivisions, partitions, planned unit developments, and other land use actions are administered by local, not state, officials.

The acknowledged local plan and its implementing ordinances set the standards for reviewing and issuing such permits. In the absence of supersiting laws, a request for a conditional use permit for a state prison, for example, would be reviewed and acted upon entirely by local officials.[2] No environmental impact statement or permit from a state agency would be required (unlike the situation in many other states).[3] The local decision would not even need to be reported to DLCD.

Those unfamiliar with the nuances of Oregon's statewide planning program often are surprised to learn that cities and counties have such authority. Oregon, after all, is widely known for the extent to which the state has asserted its power in the traditionally local process of land use planning. That reputation is well deserved: Oregon *does* protect the state's interest in land use. But the state has chosen not to assert its power directly through administrative processes such as review of routine permit applications, but through the more indirect processes of acknowledgment and periodic review.

Oregon, of course, has no "state plan." The state's policies on land use are embodied in a mosaic of 277 local comprehensive plans. Through the acknowledgment process, the Land Conservation and Development

Commission (LCDC) reviewed those plans to see that they met state standards and adequately protected the state's interests.

During acknowledgment, all state agencies were given opportunities to review and comment on the local plans. If a local plan seemed to violate state goals or rules, a state agency could object to its acknowledgment and perhaps compel changes to it. The Parks and Recreation Division of the state's Department of Transportation, for example, filed objections to several county ordinances during acknowledgment. The ordinances would have required that agency to get a conditional use permit merely to make minor improvements to an existing park. The state agency argued that local discretionary review for such actions was unduly restrictive. LCDC agreed with that reasoning, and required the local governments to allow such activities outright.[4]

In theory, the acknowledgment process enabled all the state agencies to protect their interests in that same way. Agencies that might want to site regional public facilities could, through LCDC, compel local plans to have adequate provisions for siting such facilities. In practice, however, acknowledgment sometimes failed to resolve siting issues fully, for several reasons.

First, the statewide planning goals did not address some types of regional facilities. For example, Goal 11 (Public Facilities and Services) mentions schools, solid waste disposal sites, and other "key facilities" but not correctional facilities. LCDC's powers were limited to seeing that local governments complied with the goals. LCDC therefore probably could not have required a local government to adopt provisions to facilitate the siting of a prison.

Second, the statewide goals say little about the siting process. Goal 12 (Transportation), for example, calls for each local plan to contain the provisions for a "safe, convenient, and economic transportation system." The goal, however, is silent on the role of local government in the permitting process. Should a city, for example, have the authority to deny land use permits for widening a state highway through the city center on the grounds that the local plan encourages alternative forms of transportation? Goal 12 offered no answers to such questions during acknowledgment (though it has been augmented by an innovative transportation planning rule, as discussed in chapter 6).

Third, not all state agencies actively participated in the acknowledgment process. Some agencies apparently failed to understand how their interests could be furthered through local land use plans. Others lacked the staff or expertise needed to review local plans and represent the agency's interests before LCDC. Those agencies that did not participate either could not or would not look far enough into the future to protect their interests. If a state agency did not know what types of facilities it might need or where they might be needed, it could not expect local plans and ordinances to provide for them.

Finally, some issues simply could not be anticipated during acknowledgment. In 1987, for example, Oregon's legislature adopted special supersiting legislation as part of the state's effort to bring the superconducting super collider to Oregon. Acknowledged local plans lacked provisions for siting such facilities because super colliders were unknown at the time of acknowledgment.

The process of acknowledgment gave state agencies an opportunity to shape local plans. Senate Bill 100's coordination provisions[5] offered local governments a reciprocal opportunity, through the process known as certification review.

In certification, LCDC reviews state agency programs that affect land use.[6] The object of such review is to ensure that such programs are conducted "in compliance" with the statewide planning goals and are "compatible" with acknowledged local plans. During the review, local governments may object to provisions in those programs, just as state agencies could object to local plan provisions during acknowledgment.

Such coordination generally works to enhance facility siting. It brings state agencies and local governments together to resolve policy issues. It also forces state agencies to consider land use issues, forecast the need for regional facilities, and develop siting criteria in advance of specific siting proposals.

The Supersiting System

Senate Bill 100 established an extensive system for reviewing proposals for a wide variety of land uses, including most types of regional public facilities. Oregon therefore would seem to have little need for special supersiting laws. A look at Oregon's statutes, however, reveals quite an array of such laws, each designed to reduce or eliminate local review authority over a particular type of regional public facility.

Some of the supersiting laws (ORS Chapter 469, on energy facility siting, for example) are long-term measures. They establish a separate and permanent system for the siting of an entire class of facilities. Other laws have been short-term *ad hoc* measures, intended to speed the siting of a few or even just one prison, landfill, or other facility. Examples of both types of laws are examined below.

ORS Chapter 469 assigns the responsibility for siting energy facilities such as hydroelectric dams to a state Energy Facility Siting Council. The council, whose members are appointed by the governor, has the authority to determine suitability of sites, set siting criteria, and issue site certificates. Local governments may advise on siting criteria. The council's determination that a site meets the criteria, however, prevails over state or local plans and regulations.

Senate Bill 662 in 1985 (now codified in ORS Chapter 459) declared the siting of a landfill to serve Clackamas, Marion, Multnomah, Polk, and Washington counties to be a matter of statewide concern. The counties had the primary responsibility to establish need for a disposal site, but the state could step in to site the facility. If a local government did not establish a needed site in the prescribed time, the state's Environmental Quality Commission could direct the Department of Environmental Quality (DEQ) to select a site. DEQ was required to give due consideration to local plans and ordinances. It was not required, however, to obtain permits or approvals from local government. If DEQ issued a site certificate, affected state, county, and city officials and agencies would have to approve the site. Ultimately, this legislation became moot when the Portland metropolitan area landfill was sited in Arlington, 130 miles east of Portland.

In 1987 the legislature enacted House Bill 3092, which established temporary emergency "supersiting" procedures for minimum-security correctional facilities. The law required preparation of a siting plan for an additional 1,000 prison beds. The plan set forth site selection criteria and specified general locations and capacities for each facility. Site nomination committees (consisting of representatives from the Department of Corrections, the local sheriff's office, and the local community) nominated three sites, using the criteria in the legislation and plan. A five-person Emergency Corrections Facility Siting Authority, appointed by the governor, held hearings on the nominated sites and then selected the best site. On approval by the governor, the siting authority's decision bound all affected state and local units of government, subject only to appeal to the state's supreme court.

In the 1989 legislative session, temporary prison supersiting procedures again were put into law, through House Bill 2713. This time, the purpose was to site one medium-security mega-prison. HB 2713's procedures were less open than those used for minimum-security prisons in 1987. They did not call for use of a local nominating committee and did not include any site criteria pertaining to statewide planning goals. Again, the Emergency Corrections Facility Siting Authority selected a site after holding hearings. After the governor approved a site, affected state agencies, counties, and cities were required to issue the permits needed for construction of the facility. Permit decisions were subject to review only by the state supreme court. The state prison in Ontario was sited under these provisions.

The Path Not Taken—Activities of Statewide Significance

The designers of Oregon's statewide planning program placed great emphasis on the acknowledged local plan as the controlling document for all land use decisions, but they also recognized that certain types of facilities involve issues that transcend the boundaries of the city or county where they are sited. Senate Bill 100 thus contained special provisions giving LCDC the authority to review and approve proposals for siting such facilities: sections 25-31 established a process for dealing with "activities of statewide significance."[7]

Section 25 of Senate Bill 100 said:

The following activities may be designated by the commission [LCDC] as activities of state-wide significance if the commission determines that by their nature or magnitude they should be so considered:

(a) The planning and siting of public transportation facilities.

(b) The planning and siting of public sewerage systems, water supply systems, and solid waste disposal sites and facilities.

(c) The planning and siting of public schools.

Section 26 provided that "In addition... the commission may recommend to the committee [Joint Legislative Committee on Land Use] the designation of additional activities of statewide significance."

These provisions gave LCDC the authority to create a process through which it could preempt local governments in siting regional facilities. But LCDC never used that authority. Three problems may explain the commission's reluctance.

First, LCDC lacked the staff and money necessary to undertake such a task. It devoted most of its limited resources to the task of reviewing and acknowledging 277 local plans. Second, writing new rules and siting controversial regional facilities in the face of any strong local opposition would have overextended LCDC politically. The beleaguered state commission had all it could do just to acknowledge local plans in the face of recurring statewide initiatives to do away with the fledgling program. Third, there was little evidence at that time of any need for LCDC to intervene in facility siting.

In 1981 the legislature repealed the laws that gave LCDC authority over activities of statewide significance (Oregon Laws 1981, Chapter 748, sections 1 and 12-14). The move was part of House Bill 2225's broad reshaping of the statewide planning program to "maximize efforts to complete acknowledgment" (Oregon Laws 1981, Chapter 748, preamble). The words "activities of statewide significance" no longer appear in Oregon's statewide planning laws.

Why Does Oregon Have Two Siting Processes?

Oregon has the most fully developed land use planning program in the country. That program provides the means to site most types of public facilities. Why, then, has Oregon also continued to develop *ad hoc* supersiting laws? The answer to that question lies in a complex combination of history, assumptions, and fears.

The first reason for the existence of supersiting laws is historical: some supersiting procedures were established before or in the early days of the statewide planning program. The procedure for siting energy facilities, for example, was established by laws passed in the 1970s, before the network of acknowledged local plans had been constructed.

A second reason is lack of understanding of and fear about the new statewide planning program. Particularly in the early days of the program, many people (including some legislators) did not understand how state interests were to be addressed through new and complex processes such as acknowledgment and coordination. Others simply assumed that more extensive land use planning would inevitably lead to more red tape. Both groups were inclined to believe that special legislation was needed to keep local planning and zoning from impeding development of vital facilities like landfills and prisons.[8]

A third reason might be labeled "NIMBY-itis," a fear of the not-in-my-back-yard attitude so widely reported in the 1970s and 1980s. Officials responsible for siting controversial facilities in Oregon read headlines about proposals for power plants, landfills, and prisons that were rejected in other states and probably assumed that strong measures would be needed to site such facilities in Oregon.

Another reason is the omission of some facilities from Oregon's statewide planning goals. The legislature adopted supersiting laws for prisons in 1987 and 1989, for example, partly because the goals (and hence most local plans) are silent on that topic.

A fifth reason is the variety of permits needed to site a large regional facility. Permits other than those required under planning and zoning laws are likely to be required, such as permits for new access to a public road, soil fill or removal, and sanitation, for example. Thus, even if the planning and zoning issues can be readily resolved, others may interfere with siting the facility. Supersiting legislation, however, can be written broadly enough to encompass all of the likely permits.

Another reason is simply convenience: it may be easier for an agency or interest group to get supersiting legislation for an entire class of facilities than to deal with local permit procedures in several jurisdictions. In order to get such legislation, one needs to convince a majority of legislators of only two things—that a particular type of facility is badly needed, and that local zoning may impede the facility's siting. That is likely to be a relatively easy task, since it is most often state officials who are trying to convince state lawmakers that both conditions exist.

Finally, the repeal of the provisions for activities of statewide significance left the statewide planning program without any direct means to ensure that regional public facilities could be sited. It left the partners in that program—LCDC and local governments—unable to guarantee that badly needed prisons or landfills could be built. Such uncertainty caused the advocates for certain types of facilities to seek to bypass the planning program. Their route most often took them to the legislature.

The Two Processes Begin To Converge

In 1987 contradictions and conflicts between the two siting processes intensified. In that year, the legislature passed four siting bills, all of which limited the role of local officials and plans in deciding proposals for certain types of public facilities.

❑ House Bill 3092 established a temporary supersiting process for prisons;

❑ Senate Bill 389 created a similar process for the superconducting super collider;

❑ House Bill 2936 exempted certain types of health-care facilities from local review for conditional use permits;

❑ House Bill 2884 specified that certain types of day-care facilities must be allowed outright in residential and commercial zones.

This flurry of siting legislation sparked concern on the part of local officials, many of whom had just spent the past few years getting their plans acknowledged. They argued that the state should play by its own planning rules. State officials countered that regional considerations and past difficulties in siting regional public facilities justified supersiting.

At the request of the Governor's Assistant for Natural Resources, Gail Achterman, the Department of Land Conservation and Development (DLCD) convened a temporary committee of six state and local officials to conduct that investigation during the summer of 1988.[9] The main impetus for supersiting seemed to be the fear that local planning and permit procedures would unduly hinder the siting of vital regional facilities. The committee therefore reviewed siting records for the previous decade to determine how six types of regional facilities had fared: airports; highways and other regional transportation facilities; landfills and solid-waste transfer stations; prisons and jails; child-care facilities; and residential-care facilities.

That research showed the siting record for the six types of facilities in Oregon to be quite good:

> The committee found that the typical proposal to site a public facility in Oregon has *not* encountered undue problems. Peremptory denials by local officials, neighborhood opposition, red tape, long delays, and excessive litigation have hampered a few projects. Some types of facility (notably landfills) seem more prone than others to encounter siting problems. But, the overall record for siting public facilities in Oregon is good—probably better than in most other states (DLCD 1988, p. 8).[10]

That finding led to this recommendation:

The committee recommends that supersiting legislation for any type of public facility be considered a last resort, a measure to be taken only if *all* of the following conditions are found:
1. The siting of that type of facility is a matter of statewide concern.
2. The record for siting that type of facility in Oregon shows clear evidence of significant siting problems.
3. The record shows that local planning and zoning are a principal cause of those problems.
4. The problems cannot be resolved through periodic review, state agency coordination, or other procedures under Oregon's statewide planning program.
5. Amendments to administrative rules on planning and coordination or to statewide planning goals cannot resolve the siting problems (DLCD 1988, p. 3).

The Land Conservation and Development Commission adopted the committee's report, including the above recommendation, on February 22, 1989.

The 1989 legislature appeared to heed LCDC's recommendation. It adopted only one siting bill, House Bill 2713. Although HB 2713 did create special laws to supersite one more prison, the bill also called for a transition from supersiting to siting under the provisions of the statewide planning program. The bill eliminated supersiting authority for future prisons and directed LCDC to "amend the land use planning goals and rules, adopted under ORS chapter 197, to establish streamlined siting procedures for corrections facilities" (Sections 11 and 13, Chapter 789, Oregon Laws 1989).

LCDC expanded on that legislative mandate by beginning work on a siting system that dealt not only with correctional facilities but also with other regional public facilities. At its meeting of September 21, 1990, the commission directed its staff to:

... initiate the development of a siting procedure for complex projects of regional or statewide significance. In addition to correctional facilities, the new process may address projects such as regional solid waste disposal facilities and landfills, major sewage treatment plants, dams and impoundments, and similar facilities (LCDC, minutes of the meeting of September 21, 1990).

The Department of Land Conservation and Development contracted with a group of planning consultants to propose the new facility-siting process. DLCD presented the contractors' report to the Land Conservation and Development Commission in August 1991. LCDC reviewed the report and directed its staff to prepare a draft administrative rule for hearings in 1992.

If all had gone as expected, this chapter now would be describing a new facility-siting process recently adopted by Oregon's Land Conservation and Development Commission. But early in 1992, LCDC hesitated. During a meeting on February 27, 1992, commission members agreed that the proposal had merit. They voiced concerns, however, about two issues: priorities and power.

The issue of priorities grew out of LCDC's work load. The commission faced rule making in several other crucial areas, and its budget seemed vulnerable as a result of a tax-limitation initiative passed only a few months before. Rule making for facility siting could not be considered a top-priority project.

The issue of power centered on LCDC's authority to intervene in the siting decisions made by other state agencies. The commission's chair, William Blosser, observed that LCDC might not have enough clout to establish the facility siting process through rule making; legislation might be needed.

In light of those concerns, the commission decided to ask Governor Barbara Roberts whether LCDC should proceed. In an April 27, 1992, letter to the governor, DLCD's director, Richard P. Benner, said:

> The Commission [LCDC] believes that work on facility siting is a low priority issue when compared to the need for progress on secondary and rural lands issues and urban growth management. With these considerations in mind, the Commission asked me to seek your guidance and suggestions on whether and how to proceed to complete the statutory directive to streamline correctional facility siting.

The letter went on to recommend a conservative approach: resolve the issue of siting correctional facilities by relying on existing rules and statutes for periodic review and state agency coordination. That approach was followed. LCDC turned its attention to developing new administrative rules for farm and forest lands, a Herculean task that required almost all of the agency's resources from May to December of 1992.

A Proposal for a New System

Although not adopted, the facility-siting proposal developed by DLCD and its contractors may yet serve as the basis for a new system in Oregon. It therefore merits some discussion here.

The proposal was designed to satisfy four main objectives:

1. Develop a process that could be counted on to get facilities sited.
2. Keep the decision making at a local or regional level.
3. Do not require "across the board" changes in local land use plans and regulations.
4. Integrate the facility-siting process with the present statewide planning program (Dorman, White et al. 1991).

To some extent, those objectives conflict. The proposal, however, seems to strike a balance among them, through a ten-step process outlined below:

1. An agency petitions LCDC for certificate of need. The agency submits information about the following: plans, assumptions, justification, description of alternatives (including the "no-build" option), requirements and criteria for a site, and proposed mitigation measures. The petition identifies only a general area for the project, not specific sites.

2. LCDC notifies local governments in the targeted area, recognized neighborhood groups, and other interested parties. LCDC holds a public hearing. Affected parties may propose siting criteria. They identify local plan provisions that would apply to the project.

3. LCDC approves or denies the certificate of need and the siting criteria.

4. LCDC's decision on the certificate of need may be appealed to the Court of Appeals. If LCDC approves the certificate, and if the approval withstands any appeals, issues involving need and siting criteria cannot be raised later, during selection of a specific site.

5. Local governments in the area form a regional siting council. Organization and operation of the council are established through an intergovernmental agreement (IGA). The IGA describes parties to the process, the lead agency, a schedule, procedures for hearings, staffing, mediation, and other details of the process.

6. The petitioner identifies one or more potential sites that meet the criteria set forth in the certificate of need. Local governments in the area may propose other sites. The regional siting council conducts a preliminary analysis and holds a public hearing to narrow the list of sites to at least two but no more than five.

7. **After the top candidate sites are selected, petitioner and the affected local governments must draft mitigation agreements for each site.** Affected property owners may participate. Sixty days are allowed for this step.

8. **The regional siting council (or hearings officer) holds its final public hearing.** It reviews the mitigation agreements and selects a site, using the criteria established in the first stages of the siting process.

9. **The regional council's decision is a final land use decision,** which may be appealed in accordance with Oregon's laws on such decisions. Any appeal would go to the Land Use Board of Appeals (LUBA). LUBA's decision could be appealed to the state's court of appeals and, ultimately, to the state's supreme court.

10. **The local government with jurisdiction over the selected site places a major facility overlay zone on the site.** This rezoning is a ministerial action and thus not subject to appeal. It enables the facility to be built without further review for compliance with land use standards and requirements (Dorman, White et al. 1991, pp. 44-47).

Perhaps the key feature of the siting system described above is the extent to which it separates broader policy issues from the crucial local siting question, "Should this facility go *here*?" This system has LCDC and the state agencies resolve issues about need for the facility and the criteria to be used in siting it. That reduces the number of issues to be dealt with locally, thus increasing the likelihood of efficient, equitable decision making.

The Siting Issue Remains Unresolved

In the 1980s facility siting was frequently the subject of front-page stories in Oregon newspapers. Interest in that issue has waned, however, with the construction of several prisons and regional solid-waste facilities at the end of the decade. Proponents of supersiting may argue that special legislation was responsible for those "successes." That argument, however, fails on two counts.

First, several of the sitings cannot be considered unqualified successes. The two largest facilities, the Arlington landfill and the Ontario prison, were sited far from the main regions they serve, which will add millions of dollars to the costs of operating them. Second, it is by no means clear that the facilities could not have been sited without special legislation. It was, for example, the promise of new jobs and revenues, not special laws, that enabled the Portland metropolitan area landfill to be sited in Arlington.

As Oregon moves into the 1990s, concern about facility siting has abated, at least temporarily. But the land use issues associated with this issue remain unsettled. Until they are resolved, each new proposal for a regional facility will face uncertainty about the procedures to be used in siting it. Such uncertainty will surely bring the topic of facility siting back to Oregon's legislature soon.

Notes

The views expressed in this chapter are solely those of the authors and are not official statements of policy or position on behalf of any agency.

1. As used in this chapter, the phrase "to site a facility" means to obtain the necessary land use permits and approvals to build the facility at a specific site. "Siting" thus refers to an administrative process here, not to the activities of architects or engineers in placing a structure on its site. The word "regional" is used here to mean "serving an area larger than that of the city or county where a facility is sited." The term "regional facility" thus includes large facilities, such as prisons, that serve the entire state.

2. An appeal, however, could take the decision out of the hands of local officials and into the hands of the state—most often into the hands of LUBA's three referees.

3. "Fourteen states and the Commonwealth of Puerto Rico have adopted their own environmental policy acts modeled after the National Environmental Policy Act adopted in 1969 (California, Connecticut, Hawaii, Indiana, Maryland, Massachusetts, Minnesota, Montana, New York, North Carolina, South Dakota, Virginia, Washington, and Wisconsin)" (Carpinello 1991). Other states have established special regional or state permitting procedures for regional projects. Vermont's Act 250, for example, requires all developments that involve 10 acres or more to be approved by one of nine District Environmental Commissions (Carpinello 1991).

4. See, for example, LCDC Continuance Order 82-CONT-148 (concerning Curry County's request for acknowledgment in 1982), pages 10, 11 and 18.

5. See ORS 197.180. The importance of coordination in Oregon's planning program is reflected in the title of ORS Chapter 197: "Comprehensive Land Use Planning Coordination."

6. There are more than one hundred such programs, administered by 26 state agencies.

7. Chapter 80, Oregon Laws 1973, section 3(1).

8. Those who continue to be concerned about delays in local permit processing may be overlooking several important measures that have been taken to expedite such processing. Perhaps the most significant is the 120-day limit established by state law in 1983: ORS 215.428 and ORS 227.178 require counties and cities to act on all land use decisions within 120 days of receiving an application for a permit. Any local appeals must be acted on within that period. The creation of the Land Use Board of Appeals is another measure that has expedited permit procedures. LUBA typically acts on appeals within 100 days after an appeal is filed; pre-LUBA appeals to the state's circuit courts took more than twice as long. (The figures are from the memo "Comparative Data on Performance of LUBA," February 12,

1987, from Larry Kressel, LUBA referee, to John Dubay, LUBA's chief referee.)
9. The committee's report, *Facility Siting in Oregon*, was issued by DLCD on October 24, 1988.
10 . Likewise, the 1991 report prepared by Dorman, White et al. for DLCD suggests that proposals for regional facilities in Oregon have encountered relatively few obstacles. Describing the results of a survey mailed to 277 local planners, the report (p. 26) states:

> Although a limited number of facilities have generated significant opposition, delays and expense — several of the respondents indicated that the existing land use system has worked efficiently to accommodate the majority of needed facilities . . . . In general, the local governments were very skeptical of the need for a supersiting process.

References

Carpinello, George F. 1991. *SEQRA and Local Land Use Decisionmaking*. Albany, NY: Government Law Center of Albany Law School.

Department of Land Conservation and Development (DLCD). 1988. *Facility Siting in Oregon*. Salem: DLCD.

Dorman, White & Company; Black Helterline; Faulkner/Conrad Group. 1991. *Siting Process for Facilities and Projects of Regional or Statewide Significance*. Salem: DLCD.

CHAPTER 8
Oregon Rural Land Use: Policy and Practices

James R. Pease

Since its enactment in 1973, the Oregon land use program has evolved into the most comprehensive and innovative set of rural land use policies in the United States. In 1982, the American Planning Association described it as the outstanding planning program in the nation. The conservation group Renew America called it the nation's leading growth management program in 1988, 1989, and 1990.

The components of this program are bound together by a complex framework of legislative acts, administrative rules, and judicial rulings which are administered by both state and local levels of government. As conceived in 1973, the land use program was to be carried out by collaborative working relationships between state and local governments. The state's role was envisioned as one of establishing statewide policy goals and providing oversight to ensure local government compliance with the goals. City and county governments would conduct planning studies and prepare, adopt, and administer comprehensive plans and zoning ordinances. A private organization, 1000 Friends of Oregon, was established with the backing of key political leaders and prominent citizens to monitor implementation of state policies.

While other states also enacted statewide land use legislation in the 1960s (Hawaii) and the early 1970s (Florida and Vermont), Oregon was clearly breaking new ground with the comprehensive scope of its program and the innovative requirements for citizen involvement, planning coordination, establishment of urban growth boundaries, protection of resource lands, housing policies, and other elements of the program that affected the conservation and development of land. As in other states, the appropriate roles of state and local governments in the land use program quickly became a major issue, especially in developing and implementing rural lands policy. As the Oregon program has developed and changed, state agencies and the legislature, as well as the courts, have assumed a dominant role in nearly all phases of the program, while county governments have been forced into a reactive posture of administering policies and procedures largely determined outside of their planning programs.

The substantive issues over which state and local interests have collided include resource policies affecting farm lands, forest lands, wildlife habitat, wetlands, and, recently, ground- and surface-water resources. Rural development policies have been another focus for state-local debates, specifically policies affecting urban reserve areas outside urban growth boundaries (UGBs), rural residential zones, unincorporated villages, and commercial/industrial land uses outside of UGBs. County governments would like more local control over all of these issues; state government and 1000 Friends of Oregon have advocated more policy authority at the state level. Since rural development policies are addressed in other chapters of this book, they will not be examined here. This chapter will focus on policies affecting farm and forest land.

The 1973 land use package comprised several bills which were designed to stand alone to address different concerns, but were intended to be interrelated in their implementation. For example, the Land Use Planning Act (Senate Bill 100, codified in ORS Ch. 197) provided the framework for state/local collaboration, established the Land Conservation and Development Commission (LCDC) and its administrative staff, required county governments to develop and adopt comprehensive land use plans and zoning ordinances, and empowered county governments with important coordination authority over other units of government. The County Planning Commission Act (HB 2548, codified in ORS Ch. 215.020) required each county to designate a planning director and outlined the function and membership requirements for planning commissions and the role of elected officials in land use matters. The Farm Land Tax Assessment Act (SB 101, codified in ORS Ch. 215 and ORS 308) set forth policy objectives for farm land protection, defined exclusive farm use (EFU) zones and the prescribed set of permitted and conditional uses, and set forth the conditions for obtaining property tax special assessment and deferral in EFU zones.

In 1974, LCDC adopted a set of statewide planning goals as directed by SB 100. The goal on agricultural lands (Goal 3) required each county to adopt EFU zones and incorporated the definitions and uses for EFU zones outlined in SB 101. EFU zones were to be designated by using USDA Soil Conservation Service capability classes. West of the Cascades, EFU zones were to include agricultural lands of Classes I-IV, while east of the Cascades they were to include those lands in Classes I-VI. Other lands necessary to farm operations were also to be included. Urban uses were to be contained in UGBs as defined in Goal 14. Rural lands were to be zoned as EFU, forest use (Goal 4), or "exception" areas. Exception areas were defined as unincorporated areas which were already in, or committed to, nonresource uses, such as rural residential, rural commercial, or rural industrial uses. Each exception area was required to be approved by LCDC.

Rural land use policy has been the lightning rod for the most serious challenges to Oregon's land use planning system, as well as the source of its most intractable problems. Scores of legal challenges to state and local land use policies and decisions have injected the judicial system into rural planning theory and practice. Certain judiciary decisions have created specific new planning concepts and procedures by interpreting vague statutory or administrative rule language. For example, a 1986 case (*Curry County v. 1000 Friends of Oregon*) required that LCDC differentiate between appropriate "rural" uses and "urban" uses (301 OR 447). Is a regional shopping center or automobile sales lot an urban or rural use? If urban, then it would need to be located within an UGB. In a 1985 case in Polk County, the court ruled that a farm-related dwelling permit could not be issued until a proposed farm management plan was implemented (OR LUBA 85-037). A proposed orchard, for example, would have to be established before issuance of a farm dwelling permit. In a 1989 case in Clackamas County, the Land Use Board of Appeals (LUBA) ruled that a permit for a nonfarm residence in an EFU zone can be granted only if the entire parcel is unsuitable for agriculture (OR LUBA 89-156). This ruling reversed the fourteen-year practice of allowing farmers to sell off a small nonproductive portion of a farm for a nonfarm residence. It would take a new legislative act to change the LUBA ruling.

The underlying theme in most of these cases has been state versus local control over land use decisions. The major players in the debate have been a coalition of landowners allied with land development interests, environmental protection groups (largely urban based), and county planning professionals and elected officials. The vehicles for the debates have been the courts, the legislatively created Land Use Board of Appeals, administrative rules hearings by LCDC, and the legislature, especially the Joint Legislative Committee on Land Use (JLCLU). This latter committee can, and does, exert pressure on state planners to move in certain directions when revising administrative rules. LCDC and its administrative arm, the Department of Land Conservation and Development (DLCD), have attempted to formulate policy compromises among the various interest groups, with mixed results.

While farm and forest land use policies have triggered most of the contested cases, other rural land use issues, such as adequate levels of protection in the state program for wildlife habitat, wetlands, estuaries, beaches, dunes, natural areas, scenic values, and surface- and ground-water resources have also surfaced periodically in individual cases. However, they have not yet engaged statewide attention to the same extent as have the policies affecting land divisions and building permits in farm and forest zones.[1]

A related land use issue is the state's tax deferral laws. These laws permit property tax assessment on the basis of farm use, rather than market value, for farm lands in EFU zones and for other lands in farm use, including land in rural residential zones. The difference in valuation is often on a magnitude of ten or greater, especially in the Willamette Valley and other population centers. Although it is widely recognized that the tax subsidy for noncommercial farms tends to undermine rural lands policy, by fostering demand for land partitions and homesites in resource zones, the lobby to retain the subsidy for rural hobby farms has been powerful enough to prevent legislative action.

As of 1986, all 36 counties have been certified by LCDC to be in compliance with state requirements for zoning of farm and forest lands. However, as will be discussed, the administrative rules adopted by LCDC in December 1992, together with legislation adopted by the 1993 legislature, have changed the rural land rules and will require new planning programs of many counties. Also, as each county's plan is scrutinized by LCDC under its periodic review procedures, the minimum parcel size and dwelling permit procedures in resource zones will be reexamined.

Farm and Forest Land Use Issues

Protection of the resource production base is one of the important policy goals of the 1973 land use act. Underlying these resource protection policies are several implicit assumptions about the nature of rural land markets and resource production activity:

❑ Agricultural and forest lands are scarce resources that need to be preserved for future use.

❑ Agricultural and forest harvesting activities are significant to the state's economy.

❑ Regulation of land uses is needed to prevent conversion and fragmentation of the resource base.

❑ Dwelling units in resource zones cause conflicts with resource production.

❑ Agricultural land use policy should promote larger-scale (e.g., greater than $40,000 annual gross sales) commercial agriculture.

❑ Part-time, smaller-scale (e.g., less than $40,000 annual gross sales) agriculture weakens the agricultural sector; new operations of this type should not be permitted in EFU zones.

❑ Land partitions generally weaken the commercial agriculture and forestry sectors.

❑ Without strong state guidance, local governments would issue too many partition and dwelling-unit permits, tending to undermine the commercial agriculture and forestry industries.

The database to support these assumptions is quite weak. In general, Oregon legislators and state planning officials have not fully tested the validity of widely held beliefs and assumptions underlying the land use program. For example, no study has been done on the supply and demand for farm land in the state in terms of the effects of land use or production changes on the local, state, regional, or national economy. Key assumptions about the nature and costs of conflicts between resource production and nonresource dwellings have not been supported by research findings. As will be discussed later, the studies that have been done in Oregon indicate that conflict with nearby residences has not been documented to be a significant problem for agriculture or forestry activities (Daughton 1984, McDonough 1983, Schmisseur et al. 1991).

The assumption that large-scale production units are more important to the economy than small-scale operations has also not been critically evaluated. While Census of Agriculture statewide figures clearly show that farm units larger than 100 acres and grossing over $40,000 produce most (89 percent) of the agricultural gross sales, smaller units may be important to the rural economy and social system. For example, in Deschutes County, a 1992 study found that farms grossing less than $10,000 annually purchase 27 percent of farm supplies and equipment (Pease 1992b). On a national scale, Thompson (1986) found that farms grossing under $40,000 account for over one third of farm equipment inventory. Goldschmidt (1946, 1978a, 1978b) found that smaller farms were more strongly correlated with rural community social vitality and economic activity than larger farms. Several authors in the book *Sustaining Agriculture near Cities* make the argument that small, part-time farm operations may be more resilient to changes in market conditions than are large commercial farms, thereby providing economic and social stability to rural communities (Lockeretz 1987).

The current climate of tension between county governments and the state land use agency has come about at least partly because of policy findings based on studies by the private land use monitoring group, 1000 Friends of Oregon. In various analyses of county permitting procedures, 1000 Friends has found that the majority of county land use permits in farm and forest zones did not adhere strictly to legal "findings" requirements (Liberty 1988). Under Oregon law, findings are required to link the facts of the case to the decision-making criteria, such as state laws and administrative rules. It is, to some extent, an arguable legal point when findings are "adequate." Indeed, Rohse makes the argument that the 1000 Friends studies are questionable in that the adequacy of findings was determined by the organization's staff. He also notes that their data paint a bleaker picture than warranted because they do not include denied applications. "Thus, a county that denied 99 applications . . . and approved only one, but approved it with inadequate findings, would be shown . . . with a 100 percent in the 'inadequate' column" (1986, p. 10).

In a 1987 zoning administration study, 1000 Friends of Oregon, in cooperation with the Oregon Farm Bureau Federation, reviewed a set of land use decisions in one county. They found, in an area of productive soils, cases in which productive farms had been partitioned and new dwellings approved, resulting either in new farm tracts with significantly reduced production or in land removed from farm use (Liberty 1988). Since these decisions were not randomly selected, it is not possible to use them to infer permit practices or to generalize on state policy implications. Another 1000 Friends study (Benner 1985), however, uses county permit and census data to conclude that there is what they describe as an alarming trend toward the development of hobby farms in commercial farm districts.

DLCD tracks permits in farm and forest zones and issues an annual report to the JLCLU. In response to staff concerns that certain counties (i.e., the fast-growing ones) were issuing too many dwelling and land division permits in EFU zones, LCDC included a provision in its revised administrative rules adopted in December 1992 that eleven counties, including all Willamette Valley counties, must adopt more restrictive EFU zoning provisions. These counties were selected on the basis of soil productivity and growth pressures. While DLCD data for 1990-91 permit activity indicate that, on average, western Oregon counties issued more permits than eastern Oregon counties, this pattern did not hold for all counties (see Table 1). In evaluating the effectiveness of farm zoning, it is often useful to link farm zone permit activity and acres converted to a measure of growth pressure (Mundie 1982). As Table 1 indicates, the ratio of dwelling permits on EFU land to those on non-EFU land is higher in some eastern Oregon counties than in the Willamette Valley counties. It should be noted that neither the studies by 1000 Friends of Oregon nor the DLCD reports purported to show a pattern of adverse impacts on commercial farm or forest operations by county permit activity. Spatial analysis of permit activity would clarify where dwelling permits have

Table 1. Ratio of EFU Dwelling Units/Urban Dwelling Units (August 1990-September 1991)

Willamette Valley Counties	EFU DU	Ratio of EFU DU to Urban DU	Non-Willamette Valley Counties	EFU DU	Ratio of EFU DU to Urban DU
Clackamas	52	1:28	Baker	22	1:1
Marion	51	1:18	Crook	55	1:1
Linn	27	1:4	Harney	7	2:1
Benton	32	1:5	Lake	16	1:1
Polk	31	1:4	Wheeler	10	1:1
Washington	25	1:88	Klamath	42	1:1
Yamhill	73	1:5	Umatilla	32	1:1

Source: DLCD, *1990-91 EFU Report*, Salem, Oregon, April 6, 1992.

been concentrated within the EFU zone and whether commercial resource activities were impacted. For example, mapping of nonfarm dwelling permit activity for Linn County in the Willamette Valley has shown that, while a few permits were in commercial farming areas, most of the nonfarm permits were issued in nonprime foothill areas (Sussman 1991). On the other hand, Clinton's spatial analysis of permit activity in Benton County (1993) showed that, while nonfarm dwelling permits were primarily located in the foothills, farm-related dwellings sited on small parcels were predominantly located on bottomlands and terraces, the best commercial farmland in the county. However, Clinton's analysis did not evaluate whether the small parcels were being farmed intensively or were causing problems for commercial farmers.

Secondary (small-scale) resource lands

Farm zones under the agricultural lands goal encompass 16 million acres, while the forest zones under the forest lands goal include 8.8 million acres of privately held lands. Counties are required to zone these resource lands in accordance with use and density restrictions specified in state statutes and administrative rules. The zones encompass a broad range of soil productivity characteristics and land parcelization patterns.

In response to mounting criticism that "marginal" resource lands were overprotected and that little opportunity existed for small, part-time farmers, the 1983 legislature enacted SB 273 (ORS Ch. 826), which allowed counties to designate marginal lands under specified criteria. Marginal lands would be exempt from EFU restrictions. Because counties that chose to designate marginal lands also were required to implement other more restrictive provisions for their "prime" lands, only two counties actually have used the marginal lands provisions. In 1985, the legislature directed LCDC to consider adoption of rules that would provide a method for identifying secondary resource lands and define the allowable uses of these lands. LCDC appointed a blue-ribbon committee to develop a set of recommendations for designation criteria and proposed uses in primary and secondary zones.

Several problems and policy issues emerged during this process. First was the technical problem of developing and testing a model to designate primary and secondary lands. At the minimum, the model needed to include a process to develop the database and a set of criteria for designation. In a state as geographically diverse as Oregon, the development of objective criteria with reasonable data requirements proved to be a difficult task, perhaps more because of political than technical reasons. Several efforts at model formulation have been completed since 1985, with testing done in at least eight counties, representing a cross-section of Oregon's geographic conditions.[2] The various interest groups did not come to agreement on a concept of what secondary lands should be; were they basically unproductive lands, were they unprofitable lands, or were they lands that were not "prime" or had other limitations? Hence the designation models were criticized for their results rather than their criteria and procedures. Model testing, then, became more a process of evaluating the outcome against a politically acceptable threshold of amount and type of secondary land than of evaluating and adjusting the technical criteria and procedures. Several sets of designation criteria were tested and discarded; the final version was formally adopted by LCDC in December 1992. The adopted criteria set the threshold for secondary lands (termed "small-scale") at a level very close to the line between productive and nonproductive soils. Only small tracts (i.e., less than 150 acres in eastern Oregon and less than 50 acres in western Oregon) qualified for designation as small-scale lands, whereas, in several earlier models, large tracts with unproductive soils could qualify.

A related secondary lands issue was the set of permitted uses and minimum parcel sizes appropriate for the new zones. The JLCLU and LCDC took the position that the primary zones should be more restrictive than previous EFU zones, while restrictions should be eased in the secondary zones. Farm-related dwelling permits would be limited to applicants with *bona fide* large-scale commercial farm or forestry operations in primary zones. Defining measurable standards for commercial farm and forestry units has been an important part of this issue. The lack of such standards in the original statewide Goal 3 on agricultural lands has led to much litigation and legal expense for both landowners and local governments (OR LUBA 84-006, 1984; 8 OR LUBA 128, 1983; 8 OR LUBA 201, 1983; 64 OR App 218, 1983; 68 OR App 83, 1984).

Two other important secondary lands issues have been the minimum parcel size and the property tax deferral status of secondary lands. Discussion focused on two alternative approaches for tract size: enforcement of a state-imposed 20-acre minimum or a ruling that counties would be free to develop a plan for secondary zones that could include varying minimum parcel size subzones. As to tax deferral, secondary lands could qualify by annual application and proof of farming activity, as is currently the case in non-EFU zones. Some counties have expressed concerns that many landowners may oppose rezoning to secondary lands because they will lose their "automatic" EFU tax deferral status. With enough individual opposition, it could be difficult to meet acreage block requirements to qualify for secondary zones, i.e., individual parcels must "block up" to at least 160 contiguous acres.

Not all agreed that secondary lands were necessary or desirable. Staff of 1000 Friends of Oregon argued that Oregon counties already have a more than adequate supply of lands available for rural housing and part-time farming or forestry operations through the exceptions process for rural residential zones. The problem, they have argued, has been that local officials do not apply land use laws adequately to protect the resource land base, with too many hobby farms and nonresource dwellings slipping through the system and undermining the law's intent.

Local officials and others have countered that counties have done a reasonably good job of filtering out applications for building permits in resource zones and that those improper permits that had been granted were a result of overly restrictive state regulations. Secondary lands would provide alternative sites for smaller scale and part-time resource uses, reducing local political pressures for conversion of better resource lands.

With regard to forest lands, a 1988 supreme court case (SC S33694, *1000 Friends of Oregon v. Lane County and LCDC*, March 1988) decision ruled that Goal 4 (forest lands) did not allow for dwellings in forest zones. Prior to the 1988 decision, most counties had permitted dwellings, under controlled conditions, in their forest zones and had been certified by LCDC as being in compliance with Goal 4. The court decision led to another statewide task force to recommend changes in the Goal 4 language which would permit some flexibility in allowing dwellings, while protecting the forest resource and minimizing the problems associated with forest fire control. This process was completed in January 1990 with the adoption of amendments to Goal 4 subject to any subsequent rules adopted for primary and secondary lands.

In December 1992, LCDC adopted a set of administrative rules (OAR 660-33-010 to OAR 660-33-160, OAR 660-06-070) for designation of primary and secondary (small-scale) resource lands. The new rules specified designation criteria, land uses, and minimum parcel sizes for six new zones: high-value farm lands, important farm lands, small-scale farm lands, high-value forest lands, small-scale forest lands, and mixed farm and forest lands. The new zones provided for rezoning of existing EFU and forest zones; they did not require rezoning of other lands.

High-value farm and forest zones were intended to provide a high level of protection for commercial production of farm and forest products. High-value farm lands are defined as prime, unique, Class I, or Class II soils, plus certain other lands suitable for intensive farming. High-value forest lands are defined in terms of a tract's capacity to produce 5,000 cubic feet of commercial wood fiber per year in western Oregon and 4,000 cubic feet per year in eastern Oregon. Provisions for land divisions and types of uses were also given in the administrative rules. For example, in a change to existing EFU rules, golf courses, schools, churches, private campgrounds, destination resorts, and certain other uses were no longer allowed in high-value zones. They could be permitted under certain conditions in nonhigh-value resource zones. In high-value farm or forest zones, counties could adopt LCDC-sanctioned 80-acre minimum parcel sizes (MPS) or provide documentation for a smaller MPS in high-value zones. For rangelands, the state-sanctioned MPS was 160 acres.

The new administrative rules provided that "because significant blocks of cropland exist or significant growth pressures are likely to affect cropland," the new Goals 3 and 4 rules for high-value lands were mandatory for eleven counties, including those in the Willamette Valley, Jackson County, and Hood River County. A schedule of compliance was established for the new rules, with all mandatory counties to be in compliance by 1996.

It is interesting to note the progression in terminology from "marginal" to "secondary" to "small-scale." The increasingly positive connotations of the terms may indicate the reluctance of environmental groups to give up anything of resource value. On a continuum of prime lands to unproductive lands, the threshold for small-scale clearly moved downward during the various phases of testing designation criteria.

The seven years of work on rural lands undertaken by LCDC, at the direction of the legislature, which culminated in the December 1992 administrative rules, was substantially changed by the 1993 legislature. The critics of the 1992 administrative rules joined forces with the longstanding foes of the land use program, who had gained considerable power in the 1993 House of Representatives, to loosen restrictions on dwellings in resource zones, while the majority in the Senate, as well as Governor Barbara Roberts, favored retaining strong rural land use controls. After a failed attempt to formulate an omnibus reform bill which would have substituted a wide-ranging set of rules for the 1992 administrative rules, including a much higher threshold for secondary lands, the House and Senate agreed on certain adjustments to the administrative rules in the last days of the session.

While many of the provisions of the 1992 administrative rules would stand, including the high-value designations, secondary or small-scale lands were specifically ruled out. In their place, "lot-of-record" and other rules for rural dwellings were spelled out in detail. For example, legal parcels owned prior to January 1, 1985, are now eligible for a dwelling unit permit, unless they are in a high-value zone. Nonfarm dwellings in high-value zones, which were not allowed under LCDC's 1992 administrative rules, can now be obtained with findings by a hearings officer (in the Oregon Department of Agriculture) that the parcel is 21 acres or less and is imbedded in a parcelized area as specified in the act. All other dwelling permit applicants in farm zones would have to show evidence that their parcel can support a commercial farm.

In forest zones, dwelling permits may be granted under three options: 1) if lot-of-record provisions are met for those parcels which have little productive capacity and can be served by public roads; 2) if the tract contains 160 acres in western Oregon and 240 acres in eastern Oregon if contiguous; if noncontiguous tracts are combined, the size requirement is 200 acres in western Oregon and 320 acres in eastern Oregon; or 3) if a template test is met that combines surrounding parcelization, dwelling density, and soil productivity of the site.

These dwelling permit provisions were intended to provide relief from existing rules that many local officials and landowners felt were unfair. Under previous rules, landowners had to meet stringent statutory tests for nonfarm dwellings or show evidence of a commercial farm operation. These rules had made some tracts in EFU zones unbuildable; dwelling permits in forest zones were equally difficult to obtain. HB 3661 also affirmed certain right-to-farm and right-to-forest practices by requiring written notice to purchasers of rural property that lawsuits may be limited and by limitations on local governments to restrict certain farm or forest practices.

The changes contained in LCDC's 1992 administrative rules and in the 1993 legislation are complex in their details. At this writing, neither state nor local planners have had time to fully assess the implications for rural lands policy or for landowners.

Evaluation Studies

Evaluation of the effectiveness of rural land use policies is a recent concern in the state, although several external analyses have been done of the program. Given state and local government staff and financial commitment to the program over a twenty-year period, evaluation is not only of practical interest within the state but also of academic interest throughout the country. Evaluation has proven to be no easy task, for both political and research design reasons.

Although several researchers have published papers on evaluation of the effectiveness of Oregon farm and forest land use policies and practices, the studies are largely descriptive and/or anecdotal (Leonard 1983, Pease 1992a, DeGrove 1984, DeGrove and Stroud 1987, Gustafson et al. 1982) or rely on data published by the Census of Agriculture for time interval analysis (Furuseth 1980, 1981, Daniels and Nelson 1986, Nelson 1992). For example, Daniels and Nelson (1986) concluded from a comparison of 1978 and 1982 Census of Agriculture data that Oregon led the nation in the formation of hobby farms and that this trend cast doubt on the long-term viability of commercial agriculture in the state because of land fragmentation and increasing land prices. The researchers stated that

> Both Washington and Oregon far exceeded the national average increase in the number of farms of less than fifty acres. But Oregon added 600 more of those farms than did Washington . . . These findings suggest that the Oregon program actually may be fostering the creation of small hobby farms . . . The growth in the number of hobby farms in Oregon may have negative long-term effects on the ability of commercial farms to compete for the same land, and therefore on the survival of commercial farms (p. 26).

This article has been often cited as a reason to tighten restrictions in farm and forest zones and probably contributed to the increasing mistrust of the county government permit process.

Bernhardt (1988) used special tabulations of the 1978 and 1982 Census of Agriculture to test the conclusions of Daniels and Nelson. She found that "while mid-size farms fell by 53,508 acres from 1978 to 1982, farms under 20 acres gained 22,617 acres, and farms over 320 acres gained 28,994 acres. It is possible that mid-size farms are losing more acreage to larger farms than to smaller farms and that more consolidation, rather than parcelization, is occurring" (Bernhardt 1988, p. 14). The 1982 data also show 17,740 acres of new (or newly reported) farm land, which could account for some of the acreage increase in small farms. Bernhardt also used Standard Industrial Classifications (SIC) to analyze shifts in specific types of agriculture, finding that:

> The decrease in the average size of farms has occurred primarily in groups 12 and 15 (both livestock operations), two groups which have a concentration of farms under 20 acres and which have the lowest adjusted gross sales (revenues less expenses). This may suggest that the fragmentation of agricultural land has occurred on less efficient farms (1988, p. 14).

Adjusted gross sales for all types of farms earning over $2,500 increased by nearly 20 percent between 1978 and 1982. The SIC groups with the highest adjusted gross sales showed increases in acres farmed. These trends suggest "that commercial agriculture in the Willamette Valley prospered during the study period" (Bernhardt 1988, p. 16).

The two SIC types losing the most acreage were vegetables (31,466) and livestock (20,703). While the former group lost the most acreage, it also showed an increase in gross sales of over 40 percent. Cash grains and general crop farms gained the most acreage (52,169) as well as displaying the highest increase in adjusted gross sales. A significant part of the decrease in acreage in farms grossing over $2,500 was from livestock operations, which also had low adjusted gross sales. One interpretation of these data could be that the good cropland shifted from one type of commercial agriculture to another, a common phenomenon, while any parcelization occurred primarily on grazing lands, most commonly located in foothill areas. In fact, Daniels and Nelson note that, according to local sources interviewed, most new hobby farms were located in rural residential zones or in lower quality foothill areas (1986, p. 30). These findings contradict their concern that hobby farms are undermining commercial agriculture. Bernhardt concludes, "It appears that this shift in agricultural land use benefitted the overall economic health of commercial farming during the study period" (1988, p. 16).

The trend for the 1978-82 period used by Daniels and Nelson as the basis for their concern about hobby farm proliferation is shown in Table 2. Using the same tabulations format for the period 1982-87, exactly the opposite trend is shown in Table 3.

Using Census of Agriculture data to infer land use shifts is thus fraught with uncertainties. The problems lie in the data collection methods used in different census years, the definitions used for a farm, the fact that respondents may be in rural residential as well as farm zones, and variations in the number of people responding to the questionnaire. Interviews with Census of Agriculture staff indicate that their figures are unreliable indicators of land division or building permit activity at the county level for any intercensus period (five years). Census of Agriculture data have no correlation to EFU zones nor have they been shown to have any correlation to partition or dwelling-unit permit activity. For example, a comparison between the change in the number of farms reported in the Census and number of farm-dwelling permits issued in Benton County for the period 1978-82 showed a 46 percent higher figure for the Census increase in farms than for land use permits issued during the same period (Bernhardt 1988).

Table 2. Willamette Valley Farms by Size: 1978 and 1982.

Farm Size Category	1978	1982	Change
1-49 acres	7,723	10,986	+42.3%
50-599 acres	5,414	5,076	- 6.2%
500 or more acres	745	764	+ 2.6%
All farms	13,882	16,826	+21.2%

Source: Census of Agriculture, VI, Part 37, Oregon, 1978 and 1982.

Table 3. Willamette Valley Farms by Size: 1982 and 1987.

Farm Size Category	1982	1987	Change
1-49 acres	10,986	9,900	- 9.9%
50-599 acres	5,076	4,674	- 7.9%
500 or more acres	764	791	+ 3.5%
All farms	16,826	15,365	- 8.7%

Source: Census of Agriculture, VI, Part 37, 1982 and 1987.

Nelson cites the 1987 Census of Agriculture as strongly suggesting that "Oregon's prime farmland preservation policies seem to work despite the continued proliferation of hobby farms. The conclusion is an important milestone for planning policy everywhere" (1992, p. 475). Of the 1982-87 period, he says, "Overall, Oregon lost more smaller farms but gained more larger farms than Washington or the U.S. This is limited evidence that the preservation policies discouraged proliferation of smaller farms and preserved, if not expanded, larger farms" (1992, p. 476). This conclusion is probably as unwarranted as those in the 1986 paper warning of the onslaught of hobby farms. The terms "gained" and "lost" imply a consistency in Census reporting and precision in measuring land use change that is not realistic.

Reported farm acreage over at least a ten-year period, especially acreage in farms grossing over $10,000 annually, provides a more reliable estimate of trends. As Table 4 indicates, total reported farm acreage in the Willamette Valley increased by 270,421 acres between 1978 and 1987. Land in farms grossing over $10,000 increased by 141,714 acres (about 10 percent) during the same period. While this increase in acreage could be explained by dollar inflation, it should also be noted that the Oregon economy was in severe recession during much of this period. Hobby farms do not appear to be eroding the commercial agriculture land base.

If Census of Agriculture data are an inappropriate source of information for evaluating land use change, what reliable data sources are available? Two sources have not been used to date: aerial photography and the DLCD annual reports on farm and forest permit activity. The DLCD reports, initiated in 1981, could be useful in evaluating trends if compiled in a computer database program, especially a geographic information system which would permit spatial analysis and soils/landforms analysis. To prepare and analyze the data would require research funds which the agency has not yet had available.

Table 4. Acreage in Farms in the Willamette Valley

Farm Size Category	1978	1987
All farms	1,536,792	1,807,213
Commercial farms (>$10,000 gross sales)	1,184,739	1,326,453

Source: Census of Agriculture, VI, Part 37, 1978 and 1987.

Analysis of land use change using aerial photography is a standard procedure which has not been utilized for agricultural lands in Oregon, except for a USDA Economic Research Service (ERS) national sampling study of land use change in fast-growth counties (Vesterby 1988). The statistics were published for multi-state regions and for the nation. Although some Oregon counties were included, ERS notes that the data are not reliable at the county or state level because of the limited number of sample points. The U.S. Forest Service and the Oregon Department of Forestry undertook land use change studies on forest land in 1991. The report found that "private timberland in western Oregon was not being converted to nonforest, and nonforestry land use classes were not expanding sufficiently to influence the land currently devoted to forestry" (Lettman et al. 1991, p. 11).

In 1987, LCDC provided a small grant to the Bureau of Governmental Research and Service, University of Oregon, to prepare a research design for evaluation of all elements of the state land use program. The bureau coordinated research designs by several teams of researchers, arranged for external reviews, and published a document containing the evaluation proposal (Bureau of Governmental Research and Service 1988). The proposal outlined four types of evaluation studies: 1) a political approach to determine whether citizens and elected officials are satisfied with the program; 2) an implementation approach to focus on compliance issues; 3) an impact analysis approach to measure the intended and unintended consequences of the program; and 4) a benefits approach to weigh the social and economic costs and benefits to different groups of citizens.

A key part of the proposed research design was the establishment of benchmarks or controls against which Oregon's land use policies could be evaluated. Several techniques were outlined in the proposal: 1) comparison of data before implementation of the statewide program in 1973 with 1988 figures; 2) comparison to similar jurisdictions in other states without comparable land use systems; 3) comparisons among Oregon jurisdictions; 4) an index linking land conversion rates to population growth and household formation rates; and 5) spatial analysis of permit activity as related to landforms, soils quality, and commercial resource activity. The $1.6 million proposal was submitted to the legislature but was not funded.

In 1989, the Oregon legislature appropriated $224,000 to evaluate the effectiveness of farm and forest land use policies. The JLCLU and the Emergency Board of the legislature directed LCDC to address four principal issues:

❐ What is the existing condition of Oregon's productive farm and forest lands?

❐ What are the actions and conditions that diminish the quality and quantity of farm and forest lands?

❐ What are the impacts that continuation of these actions and conditions will have on the resource lands in the future?

❐ What are the policy implications of research findings and what options are available to better protect farm and forest lands?

LCDC divided the project into three research tasks which were then put out for bids. Since LCDC formulated the research design, consultants were expected to deliver research products in accordance with contract specifications. Task One was to provide information on the condition and economic role of Oregon's farm and forest lands base (Beaton and Hibbard 1991). Task Two was to review a sample of resource lands dwelling and partition approvals made between 1985 and 1987, with some effort to evaluate management levels in 1990 (Pacific Meridian Resources 1991). Task Three was to determine the extent, types, and costs of conflicts between resource producers and nearby nonresource residents and to determine how such conflicts related to dwelling densities (Schmisseur et al. 1991).

The results were reported in June 1991. In brief, LCDC reported to the legislature that: 1) the farm and forest sectors make up about 40 to 45 percent of Oregon's economy; 2) the cumulative effect of improper land use permits is eroding the state's farm and forest lands base; 3) over half of farm dwelling and partition approvals are not directly related to commercial farm use, resulting in conflicts, higher production costs, and less land for legitimate commercial operators; and 4) additional restrictions in farm and forest zones were needed to limit new dwellings and land partitions. These new restrictions were included in revisions to Goals 3 and 4 in December 1992, providing for designation of high-value, important, and small-scale resource lands.

Economist William Fischel, in a 1991 critique sponsored by the Oregon Association of Realtors, raised questions about the study's research procedures and conclusions:

1. He estimated that agriculture and forestry together contribute only 10 percent (not 45 percent) to Oregon's economy. He attributes the gap in estimates to errors in using economic multipliers.

2. The five-year rate of prime farmland "loss" translates to 640 acres per year. This is 0.03 percent of the state's total prime farm land inventory, a relatively low rate of conversion.

3. No control was used to compare resource land management before and after land use permits were approved, making it impossible to draw defensible conclusions about productivity losses.

4. The study overemphasizes full-time commercial farmers to the exclusion of most existing farmers.

5. Complaints of dust, spray, noise, traffic, and odors from farming caused minimal changes in agricultural practices and had no effect on silvicultural practices.

6. Assuming that costs of conflicts can be extrapolated statewide, at $4 per acre, they represent little more than 1 percent of the value of annual output, a relatively minimal expense compared to other businesses.

7. The study does not demonstrate a direct relationship between higher residential densities and the frequency of costly encounters between farm and forest operators and their neighbors.

Fischel's overarching critique is that LCDC uses the study results to justify stronger state policies regarding resource development even when results are not supportive, are inconclusive, and/or are not generalizable to other areas in Oregon. While Fischel was hired by a special interest group to provide arguments for their point of view, his critique of the study emphasizes how difficult it is to obtain definitive evaluation results, even with a substantial research budget. It should be noted that contractors in the study labored under both tight time constraints and an inflexible research design.

Although Fischel is a highly regarded economist and policy analyst, his critique apparently was not taken seriously by state policy makers, perhaps because he was sponsored by a special interest group. The discrepancy between his estimate of of agriculture and forestry's contribution to Oregon's economy at 10 percent and the Task One report figure of 40 to 45 percent is certainly significant. A definitive study of this important question has not been done, according to sources in the Department of Agricultural and Resource Economics at Oregon State University (OSU), although a study is currently under way in the department. These sources estimate that Oregon's agriculture and forestry sectors may account for 10 to 20 percent of the state's economy, using value-added concepts (Cornelius 1993, Miles 1993).

The Task Three report documented widespread complaints by farmers and forest operators of trespass and litter, although widely held assumptions about restrictions on farm and forestry practices resulting from complaints of noise, odors, dust, spray, and/or slow traffic were not supported by the findings. While 70 percent of commercial farmers and ranchers surveyed reported direct conflict (litter, trespass, theft, vandalism, etc.), only 25 percent reported that these conflicts resulted in actual costs to them. For forestry, 57 percent of the sample cited direct conflicts, with 28 percent reporting costs associated with the conflicts. The costs by type of incident for replacement, repair, or change in practice were not tabulated. There was little correlation between nearby nonresource dwelling densities and these cost-related conflicts, which is consistent with findings by McDonough (1983) and Daughton (1984) for studies in Linn and Lane counties.

The Task Three findings on conflicts between resource users and rural nonresource residents are important in addressing one of the key assumptions underlying the state's rural land policies: that nonresource residents in resource zones create problems for farmers and timber harvesters and, therefore, should be sited in designated rural residential zones or inside UGBs. This assumption has been supported by some commercial farmers and timber operators. For most people, it probably seems like common sense to restrict potential conflicts which could result in costly repair, replacement, or change of farming or forestry practices.

However, the policy implications of the Task Three report are not at all clear. While trespass and litter are often problematic for nonresource rural residents as well as farmers, they were not found to have a significant negative impact on commercial agriculture. Conflicts that resulted in actual costs were not found to be significantly related to the presence of nearby residents. Some rural scholars have proposed that nonfarm residents in farm zones actually have postive effects. For example, Bryant and Russwurm (1979) proposed that, up to some threshold level of conflict, rural nonfarm residents strengthen the agricultural sector by creating a broader political base of support and by vitalizing rural communities. However, since the critical threshold levels are unknown, it would be prudent to allow nonresource dwellings in resource zones under controlled conditions, such as performance criteria related to specific site conditions and surrounding resource uses and characteristics. Such a site-specific conflict assessment model is outlined in a model resource ordinance published by OSU Extension Service (Pease and Jackson 1982) and in the development guidance system used in Hardin County, Kentucky (Planning and Development Commission 1985).

A serious oversight in the research design of Task Two was the lack of spatial analysis of permits in terms of their relationship to soils quality and the commercial resource sector and the lack of controls (e.g., comparisons to similar counties in other states) against which to measure the effectiveness of Oregon's rural land policies, although such procedures were outlined in the BGRS evaluation proposal (Bureau of Governmental Research and Service 1988). Without controls, policy implications are difficult to formulate, beyond mere review of formal compliance with state requirements. Compliance is, of course, one measure of a program's performance, but it is a weak substitute for impact analysis, since compliance mirrors a program's defects as well as strengths. The use of spatial analysis of permits and controls would have helped elucidate needed adjustments in resource lands policies.

Summary

State planners have generally interpreted evaluation studies as justifying increased restrictions on high-value resource lands, while county planners and local elected officials have argued that local programs have generally worked well in fulfilling statewide policy goals. The various studies of county permit administration by 1000 Friends of Oregon have, without question, documented cases of permits for farm dwellings and land partitions which resulted in less productive farming operations than had previously existed (Liberty 1988). The Task Two evaluation study commissioned by LCDC found that many approved farm and forest dwellings and land partitions were not being managed for resource production by property owners (Pacific Meridian Resources 1991). Clinton's (1993) spatial analysis of Benton County permits did document cases of noncommercial scale farm permits within commercial farm areas, i.e., in valley bottomlands and terraces. However, these studies were not able to establish any overall patterns or trends of negative impacts on the commercial farm and forest land base or the agricultural and forestry sectors of the economy.

The key questions of the relative importance of large (full-time, commercial) farms and small (part-time, noncommercial) farms to Oregon's rural social and economic systems and the long-term effects of nonresource dwellings on the stability of the resource land base and commercial resource operations remain largely unanswered, although assumptions about them underlie the state's rural policies. The studies cited in this text are a beginning point for understanding these complex issues. In the absence of clear answers, perhaps the current emphasis on large-scale commercial operations and restrictions on nonresource dwellings in primary farm and forest zones is the prudent approach.

Clearly, the data indicate that the commercial farm and forest base in Oregon has remained stable during the last decade. Whether this stability can be specifically attributed to the land use program or not, the land use policies for urban containment and resource land protection have undoubtedly had an important impact. The commercial agricultural sector has shown healthy increases in gross sales production, while the forest industries have shown declines for reasons unrelated to the land use program.

Some county planners have publicly expressed concerns that more restrictions on resource lands coupled with centralization at the state level of policy on rural residential, commercial, and industrial development will leave very little flexibility for local governments to cope with problems of rural growth management and economic development. In a policy position paper prepared in support of a new Goal 20, planners from Douglas County asserted that "our land use planning program has been characterized as slowly . . . choking the vitality and life out of rural Oregon. The vision for the future, the second decade of our program, must include improved opportunities for rural development. The interpretations of the statewide Planning Goals have ignored the pervasive character of Oregon's rural lifestyle. This flaw in the system is creating a stumbling block that must be removed" (Cubic 1988, p. C-1.2).

In partial agreement, James Ross, former Director of DLCD, has acknowledged that "rural development is an appropriate land use which must be explicitly recognized and addressed by the statewide goals. . . However, new rural development must not conflict with the continued protection of Oregon's primary farm and forest lands nor undermine the effectiveness of established urban growth boundaries by allowing urban levels of development in rural areas" (Ross 1988 p. G-1.5).

The rural policy issues outlined in this chapter have taken far more time and financial resources to address than anyone anticipated in 1973, when the land use laws were enacted. Additionally, state and local planners and policy makers have been led by the issues into the thorny interface between planning principles and political compromises, without much guidance from others' experience. From a researcher's perspective, the process could have been smoothed with a greater reliance on databases, policy analysis, and rigorous peer review of research findings. From a county planner's perspective, greater local flexibility and trust and a less legalistic environment could have brought the issues to faster and better resolution. From a lawyer's perspective, clearer statements of state policy objectives and more explicit procedures and definitions of terminology could have avoided much litigation and intervention by the judicial system into the planning process.

In the final analysis, however, Oregon's political leaders, planners, interest groups, and citizens have supported an exploration of new ground and a bold idea: that a state's citizens can envision and plan for their own land use future. Although its flaws may be perceived at close range, the state's ambitious attempt to control its own destiny offers a rich case history with important lessons for all students of rural land use policy.

Notes

1. The policy of siting destination resorts in resource zones had been the subject of debate but has been addressed by LCDC administrative rules. The 1992 revisions to farm and forest lands policies allow destination resorts as a permitted use in EFU zones, except on high-value lands. Policies for mineral extraction sites, especially sand and gravel pits, have created local problems in several counties. State rules require counties to identify and protect sites of high quality and quantity with little surrounding conflict for future use. Activation of extraction sites often leads to intense local protests by neighbors. While these issues are covered under existing administrative rules, the intensity of current debates indicates that the issue will probably be reexamined in the near future.
2. A detailed discussion of the secondary lands process and designation model testing is given in Pease 1990. Designation criteria testing was done in Lane, Linn, and Union counties under LCDC grants in 1988. In 1989, the Oregon legislature appropriated $242,000 for testing, which was carried out during 1989-90 in six counties: Clackamas, Coos, Deschutes, Jackson, Lane, and Union. Limited testing of new criteria was done in 1992 under LCDC grants for Jackson and Linn counties.

References

Beaton, C. R., and T. H. Hibbard. 1991. *Task One: Status of the Land Resource Base.* Salem: Farm and Forest Land Research Project, Oregon Department of Land Conservation and Development.

Benner, R. 1985. *Oregon's Farm Lands Protection Program: Is It Working?* Portland: 1000 Friends of Oregon.

Bernhardt, L. D. 1988. The Growth of Non-commercial Farming in Oregon's Willamette Valley: Assessing Impact on Commercial Agriculture. Master's Research Paper, Department of Geosciences, Oregon State University, Corvallis.

Bryant, C. R., and L. H. Russwurm. 1979. "The impact of non-farm development on agriculture: A synthesis." *Plan Canada.* 19(2): 122-139.

Bureau of Governmental Research and Service. 1988. *Evaluation Proposal for Oregon's Land Use Planning Program.* University of Oregon, Eugene.

Clinton, P. J. 1993. A Potential Growth Sales Test for Farmland: Synthesis and Application of a Rural Resource Planning Tool. Master's Research Paper, Department of Geosciences, Oregon State University, Corvallis..

Cornelius, James. 1993. Personal communication. Department of Agricultural and Resource Economics, Oregon State University, Corvallis.

Cubic, K. 1988. *Proceedings,* A Symposium on Rural Development, Sublimity, Oregon, February 11-12. Bureau of Governmental Research and Service, Eugene.

Daniels, T. L., and A. C. Nelson. 1986. "Is Oregon's Farmland Preservation Program Working?" *Journal of the American Planning Association.* 52(1):22-32.

Daughton, K. 1984. Presence of Farm and Non-Farm Produced Nuisances with the Urban Fringe of Eugene and Springfield, Oregon. Master's Research Paper, Department of Geosciences, Oregon State University, Corvallis.

DeGrove, J. M. 1984. *Land Growth and Politics.* Chicago: Planners Press.

———, and N. E. Stroud. 1987. "State Land Planning and Regulation: Innovative Roles in the 1980s and Beyond." *Land Use Law* March:3-8.

Fischel, W. A. 1991. *Much Ado About Nothing: A Critique of the 1990-91 DLCD Farm and Forest Research Project.* Dartmouth College, Hanover.

Furuseth, O. J. 1980. "The Oregon Agricultural Protection Program: A Review and Assessment." *Natural Resources Journal* 20:603-614.

———. 1981. "Update on Oregon's Agricultural Protection Program: A Review and Assessment." *Natural Resources Journal* 21:57-70.

Goldschmidt, W. 1946. *Small Business and the Community: A Study in the Central Valley of California on the Effects of Scale of Farm Operations.* Washington: US Government Printing Office.

———. 1978a. *As You Sow: Three Studies on the Social Consequences of Agribusiness.* Montclair, New Jersey: Allanheld, Osmun and Co.

———. 1978b. "Large-scale Farming and the Rural Social Structure." *Rural Sociology* 43:362-366.

Gustafson, G. C., T. L. Daniels, and R. P. Schirack. 1982."The Oregon Land Use Act." *Journal of the American Planning Association* 48(3):365-373.

Leonard, J. 1983. *Managing Oregon's Growth.* Washington: The Conservation Foundation.

Lettman, G. J., K. P. Connaughton, and N. McKay. 1991. *Private Forestry in Western Oregon: An Update on Management Practices and Land Use Changes.* Salem: Oregon Department of Forestry.

Liberty, R. 1988. *Annotated Abstract of Studies Analyzing County Administration of Oregon's Statewide Planning Conservation Goals.* Portland: 1000 Friends of Oregon.

Lockeretz, W. (ed.) 1987. *Sustaining Agriculture near Cities.* Ankeny, Iowa: Soil & Water Conservation Society.

McDonough, M. 1983. A Study of Non-Farm Dwellings in an Exclusive Farm Use Zone. Master's Research Paper, Oregon State University, Corvallis.

Miles, Stanley. 1993. Personal communication. Department of Agricultural and Resource Economics, Oregon State University, Corvallis.

Mundie, R. M. 1982. "Evaluating the Effectiveness of Local Government Farmland Protection Programs." *Geojournal* 6, 6.

Nelson, A. C. 1992. "Preserving Prime Farmland in the Face of Urbanization." *American Planning Association Journal* 58(4): 467-488.

Oregon Land Use Board of Appeals. 1983. Case No. 84-006, 1984; 8 OR LUBA 128, 1983; 8 OR LUBA 201. Salem.

Oregon Court of Appeals. 1984. Case No. 64 OR App 218, 1983; 68 OR App 83.

Pacific Meridian Resources. 1991. *Task Two: Analysis of the Relationship of Resource Dwelling and Partition Approvals Between 1985-87 and Resource Management in 1990.* Salem: Department of Land Conservation and Development, Farm and Forest Land Research Project.

Pease, J. R. 1990. "Land Use Designation in Rural Areas: An Oregon Case Study." *Journal of Soil and Water Conservation* 45, 5: 524-528.

———. 1992a. "Farm Size and Land-Use Policy: An Oregon Case Study." *Environmental Management* 15(3):337-348.

———. 1992b. Deschutes County Agricultural Resource Lands Study. Department of Geosciences, Oregon State University, Corvallis.

——— and P. L. Jackson. 1982. Model Exclusive Farm Use Zone Development Options Ordinance. Land Resource Management Program, Oregon State University, Corvallis.

Planning and Development Commission. 1985. Hardin County, Kentucky: Development Guidance System.

Rohse, M. 1986. "Farmland Protection in Oregon: Evaluating the Most Extensive Program in the Country." *Environmental Planning Quarterly* 1:7-12.

Ross, J. F. 1988. *Proceedings*, A Symposium on Rural Development, Sublimity, Oregon, February 11-12, Bureau of Governmental Research and Service, Eugene.

Schmisseur, W. E., D. Cleaves, and H. Berg. 1991. *Task Three: Survey of Farm and Forest Operators on Conflicts and Complaints.* Salem: Department of Land Conservation and Development, Farm and Forest Land Research Project.

Sussman, A. P. 1991. Data Collection and Mapping of Non-farm Related Dwelling Permits Granted in Linn County's Exclusive Farm Use Zone, July 1984-May 1990. Albany: Linn County Planning Department.

Thompson, E., Jr. 1986. *Small is Beautiful: The Importance of Small Farms in America.* Washington: American Farmland Trust.

Vesterby, M. 1988. Land Use Change in Fast-Growth Counties, Analysis of Study Methods. Staff Report No. AGES880510, Economic Research Service.

Vial, J. 1990. Spatial Distribution of EFU Permits in Benton County, Oregon, 197801990. Unpublished reoprt, Department of Geosciences, Oregon State University, Corvallis.

CHAPTER 9
Land Use Planning and the Future of Oregon's Timber Towns

Michael Hibbard

The three foundation principles of Oregon's land use planning system can be summarized as: 1) protecting rural lands; 2) minimizing the costs of new development; and 3) facilitating development within urban growth boundaries (Leonard 1983). These principles have important implications for Oregon's timber towns as they struggle to adapt to conditions that have changed radically in the twenty years since the passage of Senate Bill 100. Communities that have historically depended for their livelihoods on a resource taken from rural land are now trying to develop alternative bases for their local economies. Some of them are coping with rapid growth; others are economically stagnant.

To shed light on the role of Oregon's land use planning system during this transition, interviews were conducted during October 1991 with public officials responsible for planning in a cross-section of rural and small-town local governments. To establish a context for their observations I begin by reviewing the situation in nonmetropolitan Oregon during the early 1970s when the land use planning system was being created, and the economic and social changes that have occurred in the intervening years, especially in timber-dependent communities; I also present some background on citizens' perceptions of their communities' needs and problem-solving capacities, focusing on the role of planning. I then report the views of local government planners regarding the role of the land use planning program in addressing the issues facing their communities. Finally, I draw some broad conclusions about the role of planning in shaping the future of Oregon's small towns and rural areas.

The Twenty-Year Roller-Coaster Ride

The 1970s and '80s were among the most volatile decades in history for the communities of rural America. In the early 1970s many small towns long written off as backwaters found themselves living with new residents and new investment. But for most communities the boom did not last; from historic highs they sank to record lows in the deep recession that buffeted the country ten years later, lows from which they are still struggling to recover. This national roller-coaster ride was exactly paralleled by events in Oregon.

Goin' Up the Country

The dominant population trend in the United States during most of the twentieth century has been a rapid increase in the urban population and much slower gain in the rural population. Between 1900 and 1970, the total population of the U.S. increased by 270 percent. However, the urban population increased 440 percent while that outside of standard metropolitan statistical areas grew just 140 percent. In every decade from 1900 to 1970, the proportion of the population living in metropolitan areas increased. Suddenly, in the early 1970s, that historical pattern seemed to reverse itself. In a headline-making study, demographer Calvin Beale (1975) reported that, between 1970 and 1973, metropolitan population increased by 2.7 percent, while that in nonmetropolitan areas grew 4.2 percent. The shift was attributed to a reversal in migration patterns (Brown 1988). Changes in lifestyle preferences linked to increasing employment opportunities in rural areas led more people to move from metropolitan to nonmetropolitan counties than the reverse.

Many of Oregon's timber towns were prime targets of this trend. In some nonmetropolitan areas of the state the population grew by 4 percent a year through the first half of the 1970s (Hennigh 1981). The situation in Douglas County illustrates the local impact. Nearly half of the respondents (201 of 478) to a household survey conducted there in 1976 reported that they had moved into the county in the last five years. By far the most frequently mentioned reasons for moving to Douglas County could be called quality-of-life attractions—better neighborhoods, schools, climate, job opportunities, and the like. In general, the newcomers said they were looking for the slow pace of life and pleasant environment familiar to rural and small-town Oregonians (Umpqua Regional Council of Governments 1976). In fact, they often seemed to "out-Oregon the Oregonians" in their eagerness to become involved in the area's ongoing social processes (Hennigh 1978, p. 183).

Concern about the impact of this population growth was an important factor in the creation of Oregon's land use planning system. Oldtimers and newcomers were both anxious to preserve the rural and small-town Oregon lifestyle. According to Leonard, "the essential purpose of the Oregon land-use program is to plan for anticipated growth with minimal sacrifice of the environment and of the state's natural resource-dependent economy . . ." (1983, p. 4).

The Two Small-Town Oregons

In the 1970s, Oregon's major natural resource-based industry—timber—was concentrated in small towns. Nearly eighty of the smallest Oregon towns (population less than three thousand) had more than four-fifths of their total manufacturing workforce employed in the timber industry. In some timber towns in-migration continued through the 1980s. For example, Bend, a town that was highly dependent on timber in 1970, grew by 18.6 percent between 1980 and 1990. And in Grant's Pass, another historically timber-dependent community, the population increased by 16.3 percent in the same decade.

But even as Senate Bill 100 was being enacted the timber industry was entering a period of transformation that had unanticipated repercussions in many of the towns that land use planning was intended to protect from the consequences of the rural population turnaround. In 1979 the industry set a new aggregate production record. However, it was besieged by serious structural problems that had been obscured by the inflationary real estate boom of the latter 1970s but quickly became apparent in the steep economic downturn of the early 1980s. Production fell throughout the first half of the 1980s and many firms lost money. Vigorous competition from abroad and from other regions of the U.S., together with entry into the market of new products—such as paper pulp from eucalyptus trees and particle board as an alternative to plywood—forced the industry to cut costs and increase efficiency in order to remain competitive (Hibbard 1989a).

The Oregon timber industry returned to profitability by the mid-1980s. The 1979 production record was surpassed in both 1986 and '87, through impressive improvements in efficiency. In 1979, the industry employed 4.5 workers per million board feet of lumber produced; by 1986, the ratio had declined to 2.8 workers per million board feet. Similarly, employment in plywood manufacturing fell from 3.03 workers per million square feet of panel produced in 1979 to 2.01 workers per million square feet produced in 1986 (Ficker 1988). Put another way, the new production records were achieved with about 75 percent as many mills as were operating in 1979, and with about two-thirds as many workers.

Moreover, the reduced workforce in the small towns that depended on the timber industry were being paid less. Between May 1986 and May 1987, in the midst of the two record production years, overall hourly wages in the wood products industry fell by 3.7 percent. Hourly wages were down 4.9 percent in the logging and sawmill sectors of the industry, and 2.6 percent in plywood and veneer plants (Steward 1987). More generally, in constant dollars average hourly wages have declined each year since 1978.

In the face of this economic reality it is not surprising that two small-town Oregons have emerged. For a variety of reasons some rural communities have continued to attract in-migration. But many other places have struggled to survive in the wake of decisions that revitalized the timber industry but not the communities that depended on the industry. Between 1980 and 1990, the overall population of Oregon increased by 8.3 percent. However, the state's two major timber-producing counties barely maintained their populations. Lane County grew by 2.8 percent and Douglas County by 1.0 percent in that ten-year period. And many small timber-dependent communities actually shrank. Riddle lost 9.6 percent of its 1980 population, Drain 11.9 percent, and Oakridge 17.9 percent.

People did not move away because they wanted to. In a survey of six hundred households in six timber-dependent communities in six different Oregon counties, over half of the respondents reported experiencing economic difficulty during the period 1984-86, when the timber industry was rapidly recovering from its earlier downturn. In one-fourth of the households, at least one worker had been put on indefinite layoff; and nearly 15 percent of the households reported a member had had to move away to find work. In another random survey of three hundred Douglas

County households, conducted in the winter of 1987-88, over 20 percent of the respondents thought that their own household's financial situation was "a little worse" or "much worse" than that of others in their community (Hibbard 1989b).

A recent issue of the Portland *Oregonian* aptly if unintentionally depicted the story of the two small-town Oregons. One feature article (Church 1991) was about Brookings. The town grew by 30.0 percent between 1980 and 1990, and is struggling economically and socially to absorb its new residents. A few pages later another story told of a mill closure that will eliminate 350 jobs in Glendale, a Douglas County town with a population of only 700 (Senior 1991).

This picture of two small-town Oregons reflects a complex economic transformation that presents a major planning challenge to places that have historically depended for their livelihood on timber harvesting and processing. Some places—Bend, Grant's Pass, and Brookings are important examples—continue to attract high levels of in-migrants. They face the situation for which the Oregon land use planning system was created—how to guide growth so as to integrate the newcomers while preserving the quality of life that has attracted them. But in another, larger group of places the situation that was anticipated in 1973 has not materialized. Instead, mills are closing, jobs are disappearing, and populations are declining. The survival of these latter places depends on being able to develop a more diverse economic base while responding to the ongoing changes in their existing timber-based economy.

Local Planners' Perceptions

To understand whether and how the land use planning system is influencing the abilities of the two small-town Oregons to respond to their very different situations, a total of 21 open-ended telephone interviews was conducted during October 1991 with local officials in eleven small Oregon cities and three nonmetropolitan counties.[1] In each instance the planning director or other responsible official was interviewed regarding the effects on local economic development of various specific state-mandated land use goals; in addition, they were asked about the general effect of the land use planning system on economic development in their community and region.

The state was roughly divided into six regions: the coast, the Willamette Valley, the Columbia River, southern Oregon, central Oregon, and eastern Oregon. Within each region two incorporated cities with 1990 populations of seven thousand or less were selected, that with the largest population growth rate between 1980 and 1990, and that with the greatest rate of decline (or smallest growth).[2] Three nonmetropolitan counties were also selected, to try to understand the effects of the land use planning system on unincorporated communities and rural areas: the county with the largest rate of population growth in the decade of the 1980s, one with moderate growth, and one with substantial decline.

Any report of the respondents' perceptions of the situation in their jurisdictions must begin by emphasizing that none of the selected cities or counties is eager for change. All of the local government planners said that the deep desire to preserve the rural and small-town Oregon lifestyle of the early 1970s still prevails. Respondents from places that are experiencing rapid growth report that even those who are eager for growth do not want it to change the character of their communities. They want the land use planning system to help them maintain their quality of life.

At the same time, places that are stagnating also look to the land use planning system for help. As in the expanding communities, the residents of declining places are there because they like the community the way it is and they do not want it to change. But as community survival becomes a real question they have begun to acknowledge the necessity of planning for development and diversification.

These attitudes have had practical implications for citizen participation in local land use planning. Unsurprisingly, respondents in each jurisdiction reported that involvement was highest at the time the initial comprehensive plan was being prepared. In most places participation all but disappeared following acknowledgement, except in two situations. One is when controversy arises over specific land use changes or so-called LULUS (locally undesired land uses). The other is when proposals are made that have development implications.

In communities where the local economy is booming as well as those where it is stagnant, a pro-growth faction and an anti-change faction seem to emerge. Each side argues that its position will strengthen the community's existing quality of life. If the proposal is for new development the pro-growth faction is in support and the anti-change faction is in opposition. If the proposal is for preservation the sides switch; the anti-change faction is in support, the pro-growth faction in opposition.

The structural circumstances of small-town planning make this type of antagonistic citizen participation especially problematic. Several respondents commented on the high turnover among local planning commissioners. In other communities there is no planner on staff. In both instances the commissioners lack the necessary background of experience, training, and support to temper the (often vigorously) contentious citizen participants on either side of the issue before them. The result is often an inadequately thought-out decision that leads to a missed opportunity for a struggling community or an ill-advised development in a boomtown.

Small towns and rural communities are struggling in this perverse environment to make the land use planning system work to their advantage. Two of the system's goals seem highly salient both to communities that are experiencing growth pressure and to those that are searching for development. Goal 9 aims to improve and diversify the state's economy. Goal 8, which aims to assure that recreational facilities are sited throughout the state, is also relevant in light of the emphasis that has been placed on tourism as a vehicle for economic development.

It is here that the divergent planning needs of the two small-town and rural Oregons begin to become apparent. Communities experiencing growth report that Goals 9 and 8 have been helpful in guiding development decisions. All such communities in the sample are growing because of amenity attractions. They are tourist or retirement towns or, as in the 1970s, they are attracting in-migrants because of the local quality of life. The land use planning system has caused these places to assess the reasons for their growth and to think through their future needs. For example, two respondents described the pressures that growth exerts on agricultural and forest lands—in one case from tourism and in the other from in-migration. The land use planning system provides a forum to explicitly examine the resulting dilemmas: should agriculture and forestry be curtailed, or should expansion of tourism or in-migration be discouraged?

On the other hand, the mandate to diversify has not been very helpful to natural-resource-based communities. There is a general recognition of the need to transform the local economy but most of the respondents saw few options for their communities. Respondents mentioned unfavorable location, the lack of exploitable resources for new industries, and a shortage of suitable labor as barriers to the economic diversification of their communities. The problem of these communities is not a lack of suitable commercial and industrial properties; it is a lack of tenants for the properties.

While the utility of Goals 9 and 8 begins to illustrate the planning needs of the two small-town and rural Oregons, Goal 14 paints the fullest picture. The urban growth boundary (UGB) mandated by Goal 14 is at the heart of Oregon's land use planning system. The purpose of the UGB is to ensure "orderly and efficient transition from rural to urban land use." In essence, development is directed to sites within UGBs. How has this requirement affected small-town and rural development?

Six of the communities indicated that there is pressure to expand their UGBs. One struggling and one growing community wish to extend services to proposed recreational developments as part of their tourism strategies. Four of the growing communities report problems with housing affordability due to a lack of buildable land. They see their choices as increasing residential densities and thus changing the lifestyle patterns that have made them attractive to in-migrants, or changing their UGBs to expand the supply of buildable land. Five of the eleven communities surveyed find the UGB irrelevant as a policy instrument: each has a large supply of suitable land within its UGB and no prospects for development. The local economic development implications of Goal 14 are apparent. The UGB is a key part of each jurisdiction's plan that must be acknowledged by the state. Any change in the UGB must be renegotiated with the state.[3] This requirement can act as a brake on the sorts of hasty or ill-advised development decisions that can occur when there is great growth pressure. At the same time, as the examples above suggest, the requirement can delay or prevent the attainment of such worthy social and economic goals as the creation of jobs and affordable housing.

A fair summary of the respondents' analysis of the impact of Goal 14—and of the land use planning system as a whole—is that it has not been very effective in helping Oregon's small towns and rural communities to meet their development goals. Growth has been guided and to some extent limited in communities that are expanding rapidly, but the land use planning system has not preserved the rural and small-town Oregon lifestyle in these places. In the respondents' view, Goal 14 has introduced an urban character—high densities and escalating real estate prices—to quite small towns.

The results have also been disappointing for the large number of declining places that have looked to the land use planning system for help. The implied promise of economic health and diversification for the entire state has not materialized. Requiring small cities to locate new development within the confines of their UGB is an empty exercise when there is no development to guide and direct.

What is one to make of all this?

Land Use Planning and the Future of Oregon's Small Towns

The conventional interpretation is that Oregon's land use system reflects an urban-rural conflict (Medler and Muskatel 1979). Urban interests, economically secure and concerned with regulating rapid population growth in their communities, have prevailed over rural interests. The result is a system that seeks to restrict development to urban areas (as defined by UGBs), prohibit most development outside UGBs, and allow changes to UGBs only in response to growth pressures. Supporters of the system argue that this is good for local economies because it offers predictability to development activity and minimizes the costs of public services. Detractors hold that by limiting development the system is responsible for denying a livelihood to rural and small-town people. As we have seen, however, rural and small-town planners report a different situation. Oregon's small towns seem to be growing or stagnating for reasons that have little to do with the land use planning system.

However, places where there is growth pressure have some concerns about the land use planning system. It does guide and limit growth. But some local government planners criticize the system for imposing an "urban" urban form on growing small towns. They express concern that the higher densities and higher real estate values produced by the limiting effects of the UGB change the rural character that has made such communities so appealing to both oldtimers and newcomers. The counterargument is that the suburban sprawl that would occur if growth were not guided would have exactly the same result. It is a debate that underscores the general problem faced by planners in expanding rural communities: growth *per se* threatens to change these places. Whether guided or unguided, whether urban or suburban, growth will result in a community that is different from the one that is now attracting tourists, retirees, or other in-migrants. It is unreasonable to expect that land use planning can maintain the social and cultural character of a place when it expands by 15 or 30 percent in ten years' time. The best that rapidly growing places can do is ask what kind of new community they want under the changed circumstances, and how land use planning can help guide them there.

In the much more common situation in rural Oregon—the stagnating community—the system has had a different meaning. Citizen participation, adequate supplies of appropriate land for housing, commercial, and industrial uses, adequate recreational facilities, and urban growth boundaries have had no meaning except in rare cases when a proposed development in need of urban services—say a destination resort—could only locate outside the UGB. Generally speaking, the interests of these communities have been neither helped nor hindered by the land use planning system.[4]

In a nutshell, the economic problems of timber towns and other declining rural communities in Oregon are not a land use planning problem. Issues of timber supply, the restructuring of the timber, agriculture, and extractive industries, and changing markets for staple products will have a much greater effect than land use planning on the economic future of these places. The responses to the survey of six communities summarized above (Hibbard 1989b) indicate that people in declining rural communities understand they are faced with a socioeconomic development planning problem rather than a land use planning problem. The difference is that the land use planning system is concerned with making plans, while the focus of socioeconomic planning is on stimulating development.

The distinction was clearly made by respondents in two economically struggling communities. In each case they described the involvement of their community in local strategic planning activities sponsored by the Oregon Economic Development Department (OEDD). One respondent first characterized land use planning in her community as "an exercise in futility." The acknowledged land use plan was created "just to satisfy the legal requirement and has had negative impacts on the area, if any." At the same time, she described a volunteer citizens' committee working under the OEDD program. The committee obtained grant funds to hire an economic development specialist, drew up a local economic development plan, and is now working to implement it. In the other community the respondent told of one group of citizen volunteers who are working on a tourism plan for the area and another that is working with the Oregon Department of Transportation to improve the local road system.

Making land use plans is essential in cases where there are developments to be planned, but it is irrelevant when there is nothing on the horizon. To credit Oregon's land use planning system with bringing development to some places is as wrong-headed as blaming it for decline in other places. It is more helpful to think of the issue not in terms of urban growth versus small-town stagnation, but in terms of small-town growth versus small-town stagnation.

Conclusion

To reiterate, places are growing or stagnating for reasons that have nothing to do with the land use planning system. Land use planning neither attracts nor repels jobs and wealth. Nevertheless, the land use planning system has a continuing role to play in rural and small-town Oregon.

Small places that are faced with growth pressures may need different land use planning solutions; most importantly, people concerned with the future of rural and small-town Oregon need to understand that, while land use planning does not create growth, it can help to create agreeable physical responses to growth pressure, create the future they desire for their communities. Similarly, stagnating small towns and rural areas also need land use planning, not because it will help to solve their economic problems but because it can help to create an agreeable physical future for the community.

Notes

1. I want to acknowledge the contribution of Eric Holmes, my research assistant, to this project. Eric skillfully and patiently conducted all of the interviews reported here, using a guide that we jointly developed.
2. In two cases the second smallest city had a population greater than seven thousand, so it was selected. Also, circumstances made it impossible to complete the interview for one of the rapidly growing cities.
3. A number of small Oregon cities are completely surrounded by National Forest land or are located within the Columbia Gorge National Scenic Area. Changing the UGBs of these municipalities requires federal action as well as renegotiation with the state, since federal lands or legislation are involved.
4. Two caveats should be added to this statement. First, many respondents pointed out the undue burden that the land use planning system puts on small local governments, jurisdictions with little or no planning staff. That is an important issue, but it is beyond the scope of this research. Second, the land use planning system purports to protect the natural resource base of the state, an important source of jobs and wealth for many small towns. The effect of this on local economies—for example, on the preservation of tourist amenities or the availability of timber for struggling mills—is unclear.

References

Beale, Calvin. 1975. *The Revival of Population Growth in Nonmetropolitan America*. Washington, DC: Economic Research Service, U.S. Department of Agriculture, ERS-605.

Brown, David L. 1988. Beyond the Rural Population Turnaround: Implications for Rural Economic Development. Proceedings of the National Rural Studies Committee First Annual Meeting, Hood River, Oregon, May 24-25. Corvallis: Western Rural Development Center, Oregon State University.

Church, Foster. 1991. "Californians Spill Gold in Region." *Oregonian*. 24 November.

Ficker, Darryl. 1988. Analysis of the Regional Economic Development Strategies Program Selection Process for Lane County, Oregon. Master's Thesis, University of Oregon, Eugene.

Hennigh, Lawrence. 1978. "The Good Life and the Taxpayers' Revolt." *Rural Sociology* 43, 2: 178-190.

———. 1981. "The Anthropologist as Key Informant: Inside a Rural Oregon Town." In *Anthropologists at Home in North America*, edited by Donald A. Messerschmidt. Cambridge: Cambridge University Press.

Hibbard, Michael. 1989a. "Small Towns and Communities in the Other Oregon." In *Oregon Policy Choices 1989*, edited by Lluanna McCann. Eugene: University of Oregon Bureau of Governmental Research and Service.

———. 1989b. "Issues and Options for the Other Oregon." *Community Development Journal* 24, 2: 145-153.

Leonard, H. Jeffrey. 1983. *Managing Oregon's Growth*. Washington: The Conservation Foundation.

Medler, Jerry, and A. Muskatel. 1979. "Urban-Rural Class Conflict in Oregon Land Use Planning." *Western Political Quarterly* 32, 3: 338-349.

Senior, Jeanie. 1991. "Mill Closure Leaves Glendale Residents in Gray Mood." *Oregonian*. 24 November.

Steward, Don. 1987. "Increased Earnings—Real or an Illusion?" *Oregon Labor Trends*. (April).

Umpqua Regional Council of Governments. 1976. *Resident Housing Survey*. Roseburg, Oregon: Umpqua Council of Governments.

PART III
Perspectives and Interpretations

CHAPTER 10
The Oregon Planning Style

Carl Abbott

Addressing the 1991 convention of the Oregon Chapter of the American Planning Association, the chair of the Oregon Land Conservation and Development Commission commented that statewide planning rules allow Oregon planners to resist local development pressures and "stand up for what is good and right."

A few weeks earlier, the CEO of one of the state's wealthiest home-grown real estate development enterprises told several hundred citizens gathered to discuss transportation options for Portland that all citizens share a "moral obligation to the idea of Oregon."

These are statements to be taken seriously. Despite their lack of specific content, they are emblematic texts. The two audiences, as far as I could observe, found them perfectly reasonable contributions to discussions of urban policy. This sort of value-laden discourse, with its invocation of abstract standards of judgment, is an important window into the character of Oregon planning. In turn, this character is rooted in a broader Oregon approach to politics and public policy. There is a strong reservoir of support for land use planning in Oregon because both the concept and the processes fit with the underlying political culture and values of the state. There are certainly Oregonians who hold the individualistic view that planning impedes the full exercise of private rights or the entrepreneurial view that planning is a tool to be manipulated for private interest. Nevertheless, the majority of Oregon voters have agreed in four referenda that the state is well served by a system that defines planning as a neutral arbiter of the public interest.

This essay begins by sketching the character of Oregon's political culture and its close match with American mainstream traditions of land use regulation. It then examines the ways in which this Oregon style has manifested itself in two spheres since the beginning of the 1970s: the process of goal setting in Oregon planning, and the successful bureaucratization of planning implementation. The essay draws on specific cases which are already analyzed in the scholarly and professional literature, or can easily be documented from the public record.

❑ The establishment of the Portland Office of Neighborhood Associations (1974), the formal recognition of self-defining neighborhood associations, and the operation of neighborhood consultation and input processes (Hallman 1977, Pedersen 1979, White and Edner 1981, Cunningham and Kotler 1983, Abbott 1983, Clary 1986, Adler and Blake 1990).

❑ Planning for downtown Portland, resulting in a Downtown Plan (1972-74), a process of design review for downtown development including a Design Review Commission (1979), and a new Central City Plan in 1988 (Dotterrer 1987, Harrison 1987, Abbott 1983, 1991).

❑ Metropolitan transportation planning in the Portland area, involving the elected Metropolitan Service District and an informal council of governments in the form of a Joint Policy Advisory Committee on Transportation (Dueker et al. 1987, Adler and Edner 1990, Adler and Edner 1991).

❑ The rise and fall of the utopian community of Rajneeshpuram in central Oregon between 1981 and 1985 (Fitzgerald 1986, Abbott 1990).

❑ Evaluation of Urban Growth Boundaries as a planning and growth management tool, conducted statewide by the Oregon Department of Land Conservation and Development (1991) and in the greater Portland area by the Metropolitan Service District (1991a).

Political Culture and Planning

Each American state and community has its own style of politics and policy making. Unspoken rules, shared values, habits of behavior, and public memories of heroes and villains make it impossible to confuse Louisiana's politics with Minnesota's, New Hampshire's politics with Vermont's, or New York City's politics with Baltimore's. From John Gunther (1947) to Neal Peirce (1972), perceptive journalists have made a good living describing the differences. Cultural geographers have tried to group the individual differences into historically justifiable regions (Zelinsky 1973, Gastil 1975). Most relevant for this analysis is the argument of Daniel Elazar (1972) that American states can be categorized according to the dominance of three political cultures—traditional, individualistic, and moralistic. Moralistic communities "conceive of politics as a public activity centered on some notion of the public good and properly devoted to the advancement of the public interest. Good government, then, is measured by the degree to which it promotes the public good" (1972, pp. 96-97). The moralistic political culture, in Elazar's model, places issues ahead of individuals and accepts that government can legitimately regulate private activities such as land development for the good of the commonwealth.

Both Elazar and Ira Sharkansky (1969) place Oregon firmly among the moralists. Elazar suggests that all of the Willamette Valley cities take a moralistic approach, with weak strains of individualism intruding in Medford and Pendleton. Anticipating Raymond Gastil's idea that cultural patterns are the product of "first effective settlement," Elazar traces Oregon's moralistic approach to the dominance of New Englanders in the initial economic and political leadership of the Willamette Valley settlements, presumably reinforced by the heavy reliance on northern tier states and northern Europe for new Oregonians in the later nineteenth century. Even in the second half of the twentieth century, Oregon has been high on Episcopalians and Congregationalists, low on Baptists and Roman Catholics (Gaustad 1976). Despite substantial differences in economic interests among urban and rural sections of the state, this historic social homogeneity has remained an important factor in shaping public policies.

A more detailed look at the history of the state shows that individualistic politics contended with the moralistic style from the 1850s to the 1950s. The "official" heroes and heroines of the state's political history are people like women's rights activist Abigail Scott Duniway, progressive reformer William U'Ren, civic activist Thomas Lamb Eliot, and environmentally minded governor Oswald West. They were balanced, however, by the operators of a long-lasting Republican party machine, by well-placed practitioners of railroad and timber scams, and by Portland city administrations who were simultaneously in the hip pockets of land developers and in bed with vice industries (MacColl 1979, 1988).

The years between 1950 and 1970, however, saw the ascendancy of the moralistic side of Oregon's political culture. In very different ways, the three politicians who set the tone for public discourse in the 1950s and 1960s derived political positions from their understandings of the general public good. Wayne Morse made an extraordinary switch from

Republican to Democrat over Dwight Eisenhower's energy giveaways and was one of the earliest opponents of the Vietnam War (Unruh 1992). Mark Hatfield has based politically unpopular stands on issues such as the death penalty on his religious beliefs. Tom McCall was known for his rousing calls for protection of the Oregon environment.

McCall in particular played a central role in the development of land use planning by defining the issues in moral terms. His most famous speech in favor of state planning pointed an outraged finger at malefactors in the best style of Theodore Roosevelt. "There is a shameless threat to our environment and to the whole quality of life—the unfettered despoiling of the land," he told the legislature in January 1973. "Sagebrush subdivisions, coastal condomania, and the ravenous rampage of suburbia in the Willamette Valley all threaten to mock Oregon's status as the environmental model for the nation. . . . The interests of Oregon for today and in the future must be protected from grasping wastrels of the land" (McCall and Neal 1977, p. 196). The language is judgmental and personalized. Here is no inevitable process of land conversion driven by an impersonal market. Instead, McCall's rhetoric targeted aberrant behavior ("condo*mania*") by miscreant individuals ("grasping wastrels"). Here also is the invocation of an abstract standard of behavior that in a right world would cause the evil doers to feel *shame* for their behavior.[1]

Elazar's categorization of postwar Oregon is supported by David Klingman and William Lammers (1984), who calculated a state-by-state index of "general policy liberalism" based on levels of social service and welfare spending, anti-discrimination and consumer-protection laws and programs, date of ratification of the Equal Rights Amendment, and overall policy innovation to 1965.[2] The index runs from 1.862 (New York) to -2.061 (Mississippi). Oregon ranked sixth, behind New York, Massachusetts, New Jersey, California, and Connecticut and just ahead of Wisconsin, Minnesota, Colorado, and Michigan. The index correlated closely with Sharkansky's (1969) ranking of states along the moralistic/individualistic continuum.

Studies of metropolitan quality of life also support the description of a moralistic commonwealth. One of the earliest and most comprehensive of such studies (Liu 1975) found that very high livability for Portland and Eugene was based in part on high ranks on indicators of civic involvement such as library circulation, newspaper readership, voting turnout, and educational investment and achievement. A summary analysis of

Portland's livability has noted its participatory and issue-oriented politics (Chapman and Starker 1987). More targeted studies have found that Portland offers a civic environment that has been supportive of women's rights and economic and political opportunities (Starker and Abbott 1984, Sugarman and Straus 1988).

It is important to balance this description by pointing out two additional and superficially contradictory manifestations of Oregon's moralistic politics. One is a strain of cultural conservatism expressed in a recurring willingness to enlist the state in the enforcement of standards of personal behavior. The Ku Klux Klan in the 1920s drew strong support from native-born Protestants who wanted to force Roman Catholics to conform to a narrow conception of "Americanism." The 1990s have similarly brought vigorous efforts to put the state on record against homosexuality. It is important to understand that such efforts may be pursued without expectation of personal gain in order to advance a firmly held vision of the common good.[3]

Oregon politics are also interwoven with strong fibers of more literal status quo conservatism. Oregonians in the twentieth century have liked what they have, and they have wanted rather smugly to protect it against unwanted change. Sixty years ago, one observer compared Portland to Calvin Coolidge (Tilden 1931). Historian Gordon Dodds (1986) has characterized the postwar generation in state politics as one open to new ideas and techniques of government but committed to continuing conservatism on matters of taxation and public intervention. "Paradoxically," he has commented, "even the innovations were designed to preserve the best of the past" (p. 317). As recently as 1972, Neal Peirce (1972, p. 215) noted Portland's "anxiousness to keep things as they are." The judgment is echoed by a recent comparative analysis of economic development policy in Portland and Seattle (Abbott 1992).

Oregon's reputation for environmental protection arises from this status quo conservatism as manifested in a series of measures dating from the late 1960s to the present. Particularly important were legislation reasserting public ownership of ocean beaches, setting minimum deposits on beverage bottles and cans, creating a Willamette River Greenway, and establishing a Department of Environmental Quality. In the mid-1980s, Oregon ranked among the top six states in terms of rates of membership in a group of ten environmental organizations ranging from the Audubon Society to the Wilderness Society (Ferguson 1985). Other rankings of state environmental policy have placed Oregon sixth (*Oregonian* 1988) and first (Hall and Kerr 1991). Statewide land use planning has been an important contributor to these high comparative rankings.

Defining the Public Good: Goal Setting and Planning

In a conservative and moralistic state, land use planning has allowed Oregonians to be community minded and "good" without being revolutionary. Since the 1920s, Americans have used land use regulation to maintain established land use patterns and to protect middle- and upper-status neighborhoods from unwanted changes. As developed in the 1970s, the goals of Oregon's statewide system adapted the same impulse toward status quo conservatism to the protection of farm lands, recreational lands, and natural areas by mandating compact urbanization. Coupled with the explicit conservation goals were other goals derived from a moralistic effort to discover and realize an abstract public good. The willingness to subscribe to general goals for the public benefit can be seen in the mandate to plan land uses for efficient energy use, to preserve historic and cultural values, and to distribute low-cost housing equitably throughout the state.

It is not just the content of the statewide goals that is rooted in Oregon's political culture. The goal-setting *process* used in Oregon planning draws directly on the state's core values. It has tended to be participatory and explicitly rational. Examples include the definition of Oregon's planning goals in 1973-74 and local and statewide review of urban growth boundaries in 1990-91.

One of the key steps in the passage of Senate Bill 100 was adoption of a statement of legislative intent that state goals and guidelines were to be written by the Land Conservation and Development Commission (LCDC) only after wide public input (Abbott and Howe 1993). The new Commission devoted most of 1974 to following that directive. Arnold Cogan, the new director of the Department of Land Conservation and Development (DLCD), developed an extensive outreach program. Informational mailers with a short questionnaire went out to 100,000 randomly selected voters. Even before DLCD had found an office in Salem, it had two vans on the road fully equipped as "rolling public involvement shows" with projectors, easels, posterboard, overheads, and aspirin. Commissioners and staff met with chambers of commerce, elected officials, League of Women Voters chapters, and business clubs. They held nearly one hundred workshops around the state. The first round in the spring asked citizens to identify the physical features and qualities that were most important in their part of the state and elsewhere and to suggest means of protection. The results of the workshops and consultations were fed to technical committees that had been constituted on each goal subject. Preliminary drafts were returned to the communities in two more rounds of hearings in the late summer and fall. Ten thousand people participated directly in the drafting process through the meetings and workshops (Bureau of Governmental Research and Service 1984, Cogan 1992). Outside observers believed that the year-long process showed that "citizens of Oregon seem especially willing to dedicate their time and energy to the consideration of major public matters" (Stroud and DeGrove 1980, p. 14).

The rewrite process started with the ten general goals from the 1969 legislation and identified a number of other potential goals, including both categorical goals like energy conservation and goals targeted for sensitive areas such as the Columbia River Gorge. The public participation process helped to define and clarify the initial goals and develop detailed guidelines for local planners. It also narrowed the new candidates to a final four on housing, energy, forest lands, and citizen involvement. These were adopted along with the original ten in December 1974. In 1975 and 1976 the state added five goals that applied specifically to the Willamette River corridor and coastal areas.

The goals were constructed as an interrelated system. Many sets of planning goals suffer from being a menu of unrelated topics. At best, LCDC's famous fourteen goals are mutually supportive, so that communities pursuing one goal are likely to make simultaneous progress toward another. Special importance attaches to the set of goals that explicitly and implicitly direct metropolitan growth—Goal 3 on the preservation of farm land, Goal 5 on the preservation of open space, Goal 10 on housing opportunity, Goal 11 on orderly development of public facilities and services, Goal 13 on energy-efficient land use, and Goal 14 on the definition of urban growth boundaries.

The statewide goals have not been seriously challenged, in part because the workshop process of 1974 built a wide constituency of voters with a personal stake in the success of the program. The referendum challenges in 1976, 1978, and 1982 focused on questions of control and enforcement rather than content. The closest call in 1982 involved a ballot measure to return final decision making on land use plans to localities and retain statewide goals only as guidelines. Even the opponents of the LCDC system, in other words, argued about *how* to plan, not whether to plan, leaving the goals themselves above the political battle as examples of right thinking. During the early 1980s, when the "Sagebrush Rebellion" was mobilizing nearby states like Nevada and Utah around a radical individualist agenda, Oregon politics remained firmly centrist.

Fifteen years after adoption of the LCDC goals, both the state and the Portland metropolitan area undertook to review the operation of Urban Growth Boundaries (UGBs) as a planning tool. Defined as areas containing twenty-year supplies of developable land, UGBs obviously needed study and revision by the end of the 1980s. DLCD created a rational process grounded in an appeal to the authority of social science (in the form of ten commissioned research reports) and on the assumption that the next best steps were amenable to logical discovery (Oregon Department of Land Conservation and Development 1991, ECO Northwest 1991). The *Oregonian*, the state's largest and most influential newspaper, offered a strong endorsement when political power plays temporarily derailed the study process. Its editorial board called members of the legislature "foolhardy" and "inept" for cutting a DLCD request for additional funds to follow up and complete its 1990-91 urban growth management study, parts of which still remain unfinished (*Oregonian* 1991).

Planning for the future of the Portland metropolitan area urban growth boundary is the responsibility of the Metropolitan Service District (Metro), created in 1978 with broad planning authority to set goals through "functional plans" but not to supplant localities as a land regulator. Staff in 1988 found that Metro had no established procedures for amending the Portland area UGB, even though the state process required periodic review and anticipated incremental expansion. The agency therefore designed a classic planning process, using a policy committee to define issues while staff pulled together baseline data. Initial work was followed by consensus-based goal setting to write Regional Urban Growth Goals and Objectives (RUGGO).

A key element in the RUGGO process was the development of support through public conferences. Four hundred planners, public officials, and activist citizens turned out in 1990 and 725 in 1991. The conferences brought in outside experts like John DeGrove and Peter Calthorpe not only to tell Portlanders that the UGB had put them on the right track but also to challenge them with the information that other communities might be taking the lead from Oregon, adopting second-generation state land use programs or proactively designed compact settlements (Metropolitan Service District 1990).

The RUGGO process also involved work with local governments. Suburban cities and counties have tended to view Metro as a potential enemy ever since voters established its present form in 1978. They remember one of its two predecessor agencies (the Columbia Region Association of Governments) as a tool of Portland during the administration of Mayor Neil Goldschmidt (1973-79). Metro staff reported relatively easy interaction with suburban governments during initial stages of the process, followed by tenser discussions after the publication of draft goals (Seltzer 1991). Nevertheless, the Metro Council along with a regional policy advisory committee that included six elected city officials out of eighteen members adopted RUGGO goals in September 1991. Again in accord with a rational planning model, the next step is a Region 2040 study to define alternative transportation/land use development patterns (Metropolitan Service District 1991a, 1991b). In November 1992, tri-county voters expressed at least indirect approval of these efforts by adopting a home rule charter for Metro which includes a mandated regional planning role.

Achieving the Public Good: Bureaucracy and Planning

The rationality of goal setting in Oregon planning is complemented by successful bureaucratization of implementation. Success, in this context, can be defined as processes that regularly produce "good" planning results in accord with national professional standards, that respond to informed community consensus, and that seek to avoid the inequitable accumulation of the costs of growth and change. At its best, Oregon planning fits the regulatory (or moralistic) model that accepts the possibility and necessity of defending public interests against private power and assigns a privileged role to government. Although the scope of action for the regulatory state is theoretically unlimited, it expresses itself ideally, and sometimes in actuality, through self-controlling bureaucracies as modeled by Max Weber—neutral, rational, uninfluenced by individual status or connections. The mesh of uniform rules insulates society against special interests and pleadings (Weber 1958, Goodsell 1983, Steinberger 1985).

Bureaucracy in this conception is conservative in the same sense that Oregon's political culture is conservative. Dwight Waldo has likened it to the flywheel of a machine, providing predictability and continuity (Waldo 1971). Bureaucracy offers protection alike against antisocial behavior and disturbing social creativity. It is particularly relevant to note that the basic function of modern land use planning—to assure *predictability* in the process of land conversion and development—coincides with an essential characteristic of public bureaucracy.

When they have worked well, Oregon's planning bureaucracies have brought strong community movements into regular relationships with other economic and institutional interests. Planning in the 1980s thus helped to implement a broad community consensus forged in the 1970s by channeling high levels of public concern into accepted procedures. The same procedures can also reduce the privileges of wealth and equalize access to the public machinery of planning, helping to make outsiders into insiders. Neighborhood activists or environmental advocates can have the satisfaction of seeing their own concerns incorporated in the work programs of public agencies.

One example involves the set of neighborhood groups that appeared in Portland in the 1960s and early 1970s, some organized under the aegis of the Community Action and Model Cities programs and others springing up spontaneously in response to threats of intensified land use that would permanently alter neighborhood character. Increasing militancy and political activism led the city to explore a top-down district planning system in 1971-72. However, the choice of City Council in 1974 was establishment of an Office of Neighborhood Associations (ONA) to provide limited technical, clerical, and financial support for new or existing neighborhood groups that meet easy minimum standards of openness (Abbott 1983). The ONA system legitimized neighborhood associations, giving them automatic standing in relations with City Hall. The ONA system also set up two procedures for neighborhood input into planning decisions. Neighborhoods can play a proactive role through annual submission of neighborhood need reports that the relevant city bureaus are required to consider and respond to when developing annual budgets. Neighborhood associations are also notified of all proposed or requested zoning and land use actions affecting their neighborhood.

Two studies spanning the 1980s have shown that the neighborhood association system in Portland equalizes access to land use decisions across barriers of class and education. Charles White and Sheldon Edner (1981) found that levels of participation in neighborhood associations are roughly the same in districts of all racial and socioeconomic characteristics. Extending the analysis, Sy Adler and Gerald Blake (1990) found that the Portland system compensates for the common American tendency for homeowners and upper status persons to enjoy better access to local land use decisions than renters and lower status persons. The bureaucratized system of notification and the existence of neighborhood associations in all parts of the city equalize the likelihood that citizens will respond formally to nearby land use changes.[4]

A similar sort of equalization of haves and have-nots is seen in the implementation of downtown planning goals for Portland's historic skid road district. Portland's "North End," between West Burnside Street, North Broadway, and the Willamette River, has evolved through four stages—from a vice district to a casual labor market to a skid road dominated by alcoholics and low-income pensioners and now to a center for transients and the new homeless (Sawyer 1985). A widely accepted Downtown Plan in 1972 had held out the hope that historic preservation, commercial revitalization, and the traditional skid road could coexist north of Burnside Street. By the mid-1980s, however, it was clear that the "new homeless" of the Reagan years were far less compatible with commercial redevelopment than the "colorful" remnants of the old skid road had seemed fifteen years earlier. At the same time, a new Central City planning process (1984-88)—designed to update the Downtown Plan for a new era of growth—was targeting the entire north downtown waterfront for more intensive development.

The Portland solution to this classic land use conflict was to bypass political debate with organizational negotiation. The so-called Clark-Shiels agreement of 1987 represented a compromise between the social service community and the major property owners north of Burnside Street. The City of Portland, the Portland Development Commission, and all but one maverick social service agency signed what amounted to a peace treaty.[5] The informally negotiated agreement was adopted by the City Council in May and incorporated in the Central City Plan in November 1987. The social service groups agreed to respect a cap on the number of overnight shelter beds in the district and not to resist major public investments designed to attract private capital to the district. In return, the Development Commission agreed to move more vigorously in providing low-income and single-room-occupancy housing throughout downtown. The Clark-Shiels agreement left downtown planning in the hands of the public and nonprofit sector bureaucrats and prevented the Central City Plan from breaking down in public arguments over unmet social needs. It legitimized the social service agencies as full participants in setting public land development policy, making insiders out of potential outsiders. Both sides have proceeded with the certainty offered by a consensus plan. At the same time, the maverick agency was frozen out of the district by political muscle and bankrupted by the evaporation of contributions. No organized outsiders remain to threaten the social service/low-income housing coalition.

If the downtown planning process equalized developers and social service providers, the response to the instant city of Rajneeshpuram equalized city dwellers and small-town residents. Between 1981 and 1985, the followers of Bhagwan Shree Rajneesh built a substantial utopia on a physically isolated site in eastern Oregon. The permanent population of Rajneeshpuram reached about three thousand and the settlement incorporated as a city under Oregon law. In addition, the Rajneeshees took over the government of the nearby town of Antelope by outvoting the forty established residents and made serious threats to do the same for Wasco County (Fitzgerald 1986, Abbott 1990). The highly educated Rajneeshee leadership assumed their superiority to the rural and small-city residents of Wasco County. Operating from individualistic (and manipulative) assumptions, they treated Oregon's governmental institutions and regulatory systems as tools without inherent value, to be used when expedient and ignored when inconvenient.

Such attitudes violated the Oregon ethos that accepts the rational bureaucratic state at something like face value. The legal and bureaucratic enforcement of state land use and development regulations became an important tool by which Oregonians tried to limit the growth and impacts of Rajneeshpuram. As the Wasco County planning director later commented, "we attempted to make them follow the laws like everyone else did" (Abbott 1990, p. 95). Specifically, 1000 Friends of Oregon challenged the legality of incorporating a city on land zoned for agriculture. In the Rajneesh presentation, the town was a special effort at ecologically sound planning that Oregon should treasure as an example of enlightened development. The argument of 1000 Friends was essentially bureaucratic—that the land use system and the bad precedent were what counted. Attempts to manipulate the electoral system and the planning of criminal acts finally brought Rajneeshpuram down in the fall of 1985. However, its entanglement with building inspections, local injunctions, LCDC rulings, Land Use Board of Appeals decisions, and the Oregon appeals courts had already impeded plans for further expansion and eroded public support. By forcing the builders of Rajneeshpuram to play by the rules, the land use planning system negated much of the edge over Wasco County opponents that money and expertise might otherwise have given them.

In addition to counterbalancing the power of status and privilege, planning bureaucracies can also provide a forum for the rational consideration of the costs, benefits, and trade-offs involved in land development. A case in point is the evolution of design review in downtown Portland (Abbott 1991). Urban design emerged as an issue for Portland planning in the 1960s, tied to public concerns about urban renewal and the demolition of historic buildings. Design goals were incorporated into key planning documents in the early 1970s. Particularly important was the Downtown Plan of 1972. As already mentioned, it earned broad support by defining trade-offs among different public agencies, private interests, and competing users as part of a coherent strategy. Between 1979 and 1981, the Portland City Council converted downtown design goals into a set of routinized institutions and procedures. At the center is a Design Review Commission empowered to approve or disapprove downtown projects (with appeal to the City Council available). Design review has become an accepted part of the downtown development process. The commission usually deals with architects at early stages of a project and sometimes vetoes major proposals. One effect—and likely intent—has been to depoliticize design decisions by embedding them in a review process and insulating elected officials from direct responsibility for aesthetic choices. At the same time, the rules and guidelines let all participants debate development proposals in a common language.

Another example of diverting potential conflict into bureaucratic routines is the work of the Joint Policy Advisory Committee on

Transportation (JPACT). Until its abolition in 1978, the Columbia Region Association of Governments had met the federal requirement that local general-purpose governments participate directly in regional transportation planning. With an elected council, however, the Metropolitan Service District as created in 1978 did not meet the federal definition of a metropolitan planning organization. The response in 1979 was to create JPACT as an ad hoc council of governments. JPACT is the forum in which elected officials from local cities and counties and representatives of transportation agencies make key decisions on regional transportation policy. It is staffed by Metro's transportation planning department. The Metro Council has seldom exercised its power to reject JPACT recommendations, preferring to work toward common agreement.

The result of this double approval process has been a remarkable consensus on regional transportation strategy (Adler and Edner 1992). The first key issue was the reallocation of roughly $200 million made available by the cancellation of the 5-mile in-town Mount Hood Freeway. JPACT has since developed integrated regional highway and transit plans. It creates a level playing field and agreed rules for dealing with city-suburban and intrasuburban conflicts. To some degree, it mitigates the heavy-handed role that Portland played in transportation planning in the early and middle 1970s (Abbott 1983). Priority listings of major projects are treated as contracts. An example is light rail funding (Adler and Edner 1990). The first line was built on the east side of Portland and Multnomah County because it replaced the deprogrammed Mount Hood Freeway in the same quadrant of the metropolitan area. The second line will serve westside suburban Washington County. It was supported by the leaders of Clackamas County and by its voters in November 1990 on the assumption that a southside suburban line is securely positioned as third on the list.

Experience has taught the participants in the JPACT process that their best interests are served by maintaining a professional consensus. A united front backed by professional expertise offers advantages in competing for funds in Salem and Washington. As Sy Adler and Sheldon Edner (1992, p. 195) comment, "JPACT deals with issues in an atmosphere that presumes a continuing cooperative process. Projects are debated on their merits, and each project is given a full and fair hearing. Participants describe the process as a 'professional game' where anyone (citizen, professional, politician) who plays by the rules can have access." In this instance, the conscious depoliticizing and bureaucratizing of transportation decisions allows elected officials to minimize competition among cities and counties in favor of a metropolitan public interest.

Bureaucratic Planning in a Changing Oregon

At the trailing end of the twentieth century, Oregon land use planning is probably as good as is reasonable to expect within the rational American planning model that assumes a discoverable public interest. Oregon planning over the last two decades has been moralistic, participatory, *and* bureaucratic. The approach has worked because the state's character has made it relatively easy to find a satisfying middle ground that seems to assure the greatest good for the greatest number. In both the regional metropolis and the farming counties, the absence of a boom and bust economy has limited the political importance of wheeler-dealer land developers. The majority of residents in the Portland area share a basic vision of a relatively compact metropolis that above all else is "Not-Los-Angeles." The majority of residents statewide seem to share a parallel belief that the last, best hope is to remain "Not-California."

Despite this positive analysis, bureaucratized planning faces two interrelated challenges. One involves the need to resolve and accommodate new issues. The other stems from declining levels of public involvement and interest.

The bureaucratic strategy has worked in Oregon because planning institutions have been used to implement explicitly political decisions that involved widespread public participation and clear actions by representative bodies. Conversely, the Oregon approach can falter seriously when new, unresolved issues appear or when a substantial majority of rational people cannot agree on the common good. The state's planning institutions have just begun to face situations in which important or powerful new interests articulate positions that were not even *considered* within the public debates of the early 1970s.

An example is the case of the East Bank Freeway (I-5) in Portland. In the early 1960s, Oregon highway engineers perched Interstate 5 squarely over the east bank of the Willamette River opposite the central business district, creating bottleneck curves and visual blight. In the mid-1980s, a combination of neighborhood groups, transit advocates, and design professionals mounted a campaign to undo the work of the engineers by moving the freeway several blocks off the river into a thriving industrial loft district, presumably to free land for a waterfront park or visually attractive redevelopment. Politicians were surprised both by the emergence of the issue and by the rapid accumulation of articulate support. At the same time, they feared that acceptance of a new and expensive ($50-100 million) highway improvement would unravel the carefully constructed regional consensus on transportation investments. Portland officials also had to balance aesthetic and design values against possible disruption of the city's economic base. Feelings ran high enough on both sides that the issue was excluded from the Central City Plan in order not to hold it hostage. A series of ad hoc study committees in 1987-88 failed to reach agreement and the City Council voted 4-1 for the status quo in March 1989. It is fair to say that each side believed the other to have ignored the obvious public good in favor of a narrow agenda. Indeed, continued dissatisfaction and agitation by the advocates of relocation prompted the City Council to reopen the debate in 1993 by appointing a new citizen advisory committee to try to sort out the unreconciled positions.

Another politicized issue has involved the question of defining "secondary lands." As James Pease has noted in Chapter 8, the legislature in both 1985 and 1987 asked LCDC to devise a way to allocate rural lands among those capable of sustaining profitable agriculture or forestry ("primary lands") and those lower quality secondary lands that might properly be opened to some level of more intense development. In response, LCDC tied an easing of restrictions on secondary lands to tighter limits on primary lands. However, what looked like a good planning compromise proved highly controversial. Perhaps trying to reconstruct the success of its 1974 goal-setting process, LCDC held nearly a score of hearings around the state but still failed to find common ground among urban environmentalists, commercial farmers, and owners of wood lots and forest land. In 1991 the issue passed to the legislature, which also failed to resolve competing bills. While citizens and politicians could not agree whether the objective was to find as much developable land as reasonable or as little as possible, LCDC responded to the legislative request by adopting what it thought would be acceptable compromise rules in December 1992. In fact, its actions satisfied neither side in the debate—a good example of the failure of the bureaucratic approach in the absence of political consensus. In turn, the 1993 legislature reopened the debate and agreed on its own replacement for the LCDC rules by passing House Bill 3661 in the last week of its session, potentially ending nearly a decade of political controversy over what had initially been viewed as a technical land use issue.

These examples are indicative of the second problem. The very nature of bureaucratization involves the substitution of process for substantive debate, the replacement of enthusiasm by technical routine. Direct citizen participation is likely to be replaced with indirect participation through the professional staff of organizations and interest groups.[6] With the "big decisions" already made, LCDC's formal statewide advisory committee on citizen involvement has had little work and limited impact over the last decade (Abbott and Howe 1993). Indeed, it might be argued that a functioning bureaucracy has greater need for something like performance auditing than for citizen involvement in the style of 1974.

But there is now an entire generation of Oregonians who did not participate in the political battles of 1966-74, the statewide goal-setting processes of 1974-76, or local comprehensive planning in the later 1970s. Citizens new to planning issues come face to face with state guidelines and local comprehensive plans as parts of a system of abstract rules interpreted by specialist lawyers and planners. Their response is often the loud complaint that the system fails to recognize *their* interests or concerns. Such participation by protest can come from either side of the political spectrum—from rural property owners unhappy about development restrictions or from Portland community activists angered by neighborhood gentrification. In either case, it forces planning questions out of the bureaucracy and back into direct political discussion.

The crisis for Oregon's planning system may come toward the end of this decade, when uncommitted or skeptical voters are matched with a new set of political leaders who also have not experienced the earlier public involvement in land use planning. We have a preview of the results at the small scale with the Eastbank Freeway and the large scale with secondary lands. As Oregonians struggle with the implications of industrial transition and globalization, they will need to shape new definitions of the public good that may well create new planning tasks. It will be the responsibility of citizens and elected officials to continue the Oregon tradition of seeking the common ground. It will be the responsibility of planning professionals to respond with planning tools that match the effectiveness of those created in the 1970s and 1980s.

Notes

1. Oregon's taste for politicians who claim to put principle above party was reflected in Ross Perot's 25 percent share of the 1992 presidential vote in the state. Perot, of course, presented himself as an outsider who spoke for the broad public interest in contrast to the special interest agendas espoused by self-serving professional politicians.
2. Examining the timing of adoption of 88 policies and programs, Walker (1969) found that Oregon ranked eighth as an innovator among 48 states.
3. A strongly anti-homosexual constitutional amendment drew 43 percent of the vote in November 1992. Oregon courts in December 1992 overturned a successful 1988 ballot measure that had rescinded Governor Neil Goldschmidt's executive order barring discrimination against homosexuals in state employment.
4. At the start of the 1990s, nearly every neighborhood in Portland wanted its own neighborhood plan. Neighborhoods actively seek city-funded plans staffed by the Planning Bureau. They work for designation as historic districts. They try to produce their own plans in the absence of city assistance. This very enthusiasm, however, raises points of serious potential conflict. Schooled in the contingencies of urban development, professional planners are likely to see neighborhood plans as flexible guides to controlled change. Neighborhoods are likely to see them as contracts that define a status to be preserved and to so remember neighborhood plans made up to twenty years ago. The city's practical response has been to opt for a complete series of "district" or "community" plans that will each cover a cluster of neighborhoods.
5. The memorandum was negotiated by Don Clark, a former Multnomah County Commissioner and executive director of Central City Concern, a large social service agency, and by Roger Shiels, a consultant with a reputation for facilitating major projects. Clark signed on behalf of the "social service community," Shiels on behalf of "the business community," particularly the larger property owners of the Burnside district who would benefit from a northward extension of the Transit Mall and similar public improvements.
6. For example, the establishment of the Columbia River Gorge National Scenic Area in 1986 has replaced passionate environmental lobbying with complex but substantially sympathetic bureaucratic processes that pay close attention to the expert groups (Adler 1990).

References

Abbott, Carl. 1983. *Portland: Planning, Politics, and Growth in a Twentieth Century City.* Lincoln: University of Nebraska Press.

———. 1990. "Utopia and Bureaucracy: The Fall of Rajneeshpuram, Oregon." *Pacific Historical Review* 59, 1: 77- 103.

———. 1991. "Urban Design in Portland, Oregon, as Policy and Process: 1960-1989." *Planning Perspectives* 6, 1: 1-18.

———. 1992. "Regional City and Network City: Portland and Seattle in the Twentieth Century." *Western Historical Quarterly* 23, 3: 293-322.

———, and Deborah Howe. 1993. "The Politics of Land-Use Law in Oregon: Senate Bill 100, Twenty Years After." *Oregon Historical Quarterly* 94, 1: 5-35.

Adler, Sy. 1990. "Environmental Movement Politics, Mandates to Plan, and Professional Planners: The Dialectics of Discretion in Planning Practice." *Journal of Architectural and Planning Research* 7, 4: 315-29.

———, and Gerald Blake. 1990. "The Effects of a Formal Citizen Participation Program on Involvement in the Planning Process: A Case Study of Portland, Oregon." *State and Local Government Review* 22, 1: 37-43.

———, and Sheldon Edner. 1990. "Governing and Managing Multimodal Regional Transit Agencies in a Multicentric Era." In *Public Policy and Transit System Management*, edited by George Guess. Westport, CT: Greenwood Press.

———. 1992. "Challenges Confronting Metropolitan Portland's Transportation Decision System." *Transportation Research Record*, No. 1364. Washington: National Academy Press.

Bureau of Governmental Research and Service. 1984. *Guide to Local Planning and Development.* Eugene: University of Oregon.

Chapman, Nancy, and Joan Starker. 1987. "Portland: The Most Livable City?" In *Portland's Changing Landscape*, edited by Larry Price. Portland: Portland State University and Association of American Geographers.

Clary, Bruce. 1986. *A Framework for Citizen Participation: Portland's Office of Neighborhood Associations.* Washington: International City Management Association.

Cogan, Arnold. 1992. Interview by author, 17 December.

Cunningham, William V., and Milton Kotler. 1983. *Building Neighborhood Organizations.* Notre Dame, IN: Notre Dame University Press.

Dodds, Gordon. 1986. *The American Northwest: A History of Oregon and Washington.* Arlington Heights, IL: Forum Press.

Dotterrer, Steve. 1987. "Changes in Downtown Portland." In *Portland's Changing Landscape*, edited by Larry Price. Portland: Portland State University and Association of American Geographers.

Dueker, Kenneth, Sheldon Edner, and William Rabiega. 1987. "Transportation Planning in the Portland Metropolitan Area." In *Portland's Changing Landscape*, edited by Larry Price. Portland: Portland State University and Association of American Geographers.

ECO Northwest. 1991. *Urban Growth Management Study: Case Studies Report*. Salem: Department of Land Conservation and Development.

Elazar, Daniel. 1972. *American Federalism: A View from the States*. New York: Thomas Y. Crowell.

Ferguson, Kathleen A. 1985. Toward a Geography of Environmentalism in the United States. Unpublished Master's Thesis, California State University, Hayward.

Fitzgerald, Frances. 1986. *Cities on a Hill*. New York: Simon and Schuster.

Gastil, Raymond. 1975. *Cultural Regions of the United States*. Seattle: University of Washington Press.

Gaustad, Edwin S. 1976. *Historical Atlas of Religion in America*. New York: Harper and Row.

Goodsell, Charles. 1983. *The Case for Bureaucracy*. Chatham, NJ: Chatham House Publishers.

Gunther, John. 1947. *Inside USA*. New York: Harper and Brothers.

Hall, Bob, and Mary Lee Kerr. 1991. *The 1991-1992 Green Index: A State by State Guide to the Nation's Environmental Health*. Washington: Island Press.

Hallman, Howard. 1977. *The Organization and Operation of Neighborhood Councils*. New York: Praeger.

Harrison, Michael. 1987. "Promoting the Urban Experience in Portland, Oregon." In *Public Streets for Public Use*, edited by Anne Moudon. New York: Van Nostrand Reinhold Co.

Klingman, David, and William W. Lammers. 1984. "The 'General Policy Liberalism' Factor in American State Politics." *American Journal of Political Science* 28, 3: 598-610.

Liu, Ben-Chieh. 1975. *Quality of Life Indicators in U. S. Metropolitan Areas, 1970: A Comprehensive Assessment*. Washington: U. S. Environmental Protection Agency.

McCall, Tom, and Steve Neal. 1977. *Tom McCall: Maverick*. Portland: Binford and Mort.

MacColl, E. Kimbark. 1979. *The Growth of a City: Power and Politics in Portland, Oregon, 1915-1950*. Portland: Georgian Press.

———. 1988. *Merchants, Money, and Power: The Portland Establishment, 1843-1913*. Portland: Georgian Press.

Metropolitan Service District. 1990. *Planning a Livable Future: Growth Strategies for the 21st Century: Proceedings of the 1990 Regional Growth Conference.* Portland: Metropolitan Service District.

———. 1991a. *Building a Livable Future: The Revised Regional Urban Growth Goals and Objectives.* Portland: Metropolitan Service District.

———. 1991b. *Metro Planning News.* June, September.

Oregon Department of Land Conservation and Development. 1991. *Urban Growth Management Study: Summary Report.* Salem: Department of Land Conservation and Development.

Oregonian. 1988. "Oregon Ranks in Top Ten." 25 February.

———. 1991. "Give Planners the Tools." 1 October.

Pedersen, Mary. 1979. *A Five Year Evaluation of Portland's Neighborhood and Citizen Participation Program.* Portland: City of Portland, Office of Neighborhood Associations.

Peirce, Neal. 1972. *The Pacific States of America.* New York: W. W. Norton and Co.

Sawyer, Chris. 1985. From Whitechapel to Old Town: The Life and Death of the Skid Road District, Portland, Oregon. Unpublished Ph.D. Dissertation, Portland State University.

Seltzer, Ethan. 1991. Interview by author, 4 April.

Sharkansky, Ira. 1969. "The Utility of Elazar's Political Culture." *Polity* 2, 1: 66-83.

Starker, Joan, and Carl Abbott. 1984. "The Fourteen Best Cities for Single Women." *Ms.* 13, 5: 129-32.

Steinberger, Peter. 1985. *Ideology and the Urban Crisis.* Albany: State University of New York Press.

Stroud, Nancy, and John DeGrove. 1980. *Oregon's State Urban Strategy.* Washington: National Academy of Public Administration.

Sugarman, David B., and Murray A. Straus. 1988. "Indicators of Gender Equality for American States and Regions." *Social Indicators Research* 20: 229-70.

Tilden, Freeman. 1931. "Portland, Oregon: Yankee Prudence on the West Coast." *World's Work* 60, 10: 34-40.

Unruh, G. Q. 1992. "Republican Apostate: Senator Wayne L. Morse and His Quest for Independent Liberalism." *Pacific Northwest Quarterly* 82, 3: 82-91.

Walker, Jack. 1969. "Diffusion of Innovation among the American States." *American Political Science Review* 63, 3: 880-89.

Waldo, Dwight. 1971. *Public Administration in a Time of Turbulence.* Scranton, PA: Chandler Publishing Co.

Weber, Max. 1958. *From Max Weber: Essays in Sociology.* Edited by Hans Gerth and C. Wright Mills. New York: Oxford University Press.

White, Charles, and Sheldon Edner. 1981. "Participation in Neighborhood Organizations." *Journal of Community Action* 1, 1 : 48-52.

Zelinsky, Wilbur. 1973. *The Cultural Geography of the United States.* Englewood Cliffs, NJ: Prentice-Hall.

CHAPTER 11
Following in Oregon's Footsteps: The Impact of Oregon's Growth Management Strategy on Other States

John M. DeGrove

The rise of the environmental movement in the United States after World War II, reaching a peak in the early 1970s, led to the adoption of a number of state land use laws we would now label as growth management systems. Beginning with Vermont in 1970 and ending with Hawaii in 1978, seven states adopted more or less comprehensive systems for reallocating responsibility for land and growth management within and among state, regional, and local levels. All of these laws resulted in some sharing of authority and responsibility by local governments with state and regional levels of government. The California and North Carolina laws were confined to coastal areas; the Vermont and Florida approach involved limited state/regional roles for land use decisions of greater than local impact; the Hawaii system started as a comprehensive approach but was substantially weakened by later changes in the law; and the Colorado system was limited in scope and became weaker over time in the face of a hostile legislature and adverse court rulings. Only Oregon among these seven states adopted and implemented a comprehensive system containing all but one of the key concepts of the growth management systems of the 1980s. That exception was the failure to address the issue of concurrency, the requirement that infrastructure already be in place to accommodate the impacts of new development. Because of the scope of Oregon's program, as well as its sustained political support in the face of multiple challenges, Oregon has influenced heavily and in more or less direct ways the content of growth management systems adopted during the 1980s (DeGrove 1984).

The Key Concepts of the Oregon Program

Seven concepts from the Oregon program have influenced other state growth management systems: 1) consistency; 2) urban growth boundaries; 3) the protection of farm and forest lands; 4) a positive affordable-housing strategy; 5) a focus on economic development; 6) mandates leading to certainty and timeliness by local government in their planning and regulation systems; and 7) the establishment of a watchdog group to support the survival and timely implementation of the system.

Oregon's impacts on Florida

After an initial effort at crafting an effective growth strategy in the first half of the 1970s, growing frustration with the gaps in and other weaknesses of the system caused Florida to return to the drawing board in the mid-1980s for a second try at putting in place a strong capacity for managing the state's massive growth pressures. The centerpiece of Florida's reexamination of its growth management systems was a broad-based blue ribbon commission titled the Second Environmental Land Management Study Committee (ELMS II). Over an 18-month period from 1982 to 1984, the group heard testimony from knowledgeable people from across the nation about growth management experiences from other states.

One of the best received of this group was Henry Richmond, executive director of 1000 Friends of Oregon, whose testimony focused on how the Oregon system's goals and policies worked together through the requirement that local plans be "acknowledged" as consistent with nineteen state goals and policies. These goals, among other things, required compact urban development patterns that protected important farm and forest land and promoted affordable housing through mandated density increases inside urban growth boundaries. This testimony and follow-up communications between 1000 Friends and ELMS II were reinforced by the author, a member of ELMS II and one of the architects of the growth management legislation of 1985. Dr. John DeGrove had made an in-depth study of the Oregon system (DeGrove 1984). When he was named secretary of the Department of Community Affairs (the state land planning agency), he was in a key position to push for inclusion in the Florida law of key concepts pioneered by Oregon. While the give and take of legislature politics in Florida weakened some of the key Oregon goals and policies, elements of all of them were included in Florida's two major legislative acts approved in 1985: the Omnibus Growth Management Act and the State Comprehensive Planning Act. Florida's consistency requirement was, if anything, more tightly drawn than Oregon's, and the process moved considerably faster. Florida's consistency requirement called for each complete plan to be presented for state review at a specific date during the four-year span from mid-1988 to mid-1992, not one element at a time with constantly extended time frames (DeGrove 1991). As of July 1993, more than 90 percent of Florida cities and counties were in compliance or had compliance agreements in place.

The architects of Florida's legislation struggled mightily to include strong measures against urban sprawl in the law, and drew on the Oregon record in the successful implementation of an urban growth boundary (UGB) requirement. Resistance to such a clear requirement was strong in the Florida legislature, however, and no absolute requirement was included in the law. Nevertheless, some key goals and policies were included in the legislation, including the separation of rural and urban uses, the concentration of development where infrastructure was in place, and, in a 1986 addition to the law, an admonition to discourage urban sprawl. An aggressive application of these policies added requirements for compact urban development to local government plan preparation. These requirements utilized such concepts as urban service areas as rough surrogates for Oregon's UGBs, with some of the implementation ideas drawing directly from the Oregon experience. For example, Florida borrowed from Oregon a target limit on the amount of land available for development, setting the target at the amount needed to accommodate no more than 125 per cent of the projected population growth. The application of this policy in Florida has been extremely controversial, but it has been upheld so far through decisions by administrative hearing officers and the courts. Still not so powerful a policy as Oregon's UGBs, it is a large step in that direction that continues to gain strength (Florida Department of Community Affairs 1989).

While the protection of farm and forest land, affordable housing requirements, and economic development strategies in Florida's law are not as strong as those policies in Oregon, they have also gained strength in the implementation of the law to date, and the reference point for stronger programs in these areas has been Oregon's experience. The executive director of 1000 Friends of Oregon and other Oregonians have returned to Florida in recent years to explain the close link between UGBs, the protection of farm and forest land, and affordable housing. Florida's review of local plans with regard to these concepts has clearly been influenced by the Oregon experience. Local plans that invite sprawl at the expense of farm and forest land and natural resources by placing such land in relatively high-density categories (from one to three dwelling units per acre) have been found inconsistent with the relevant state and regional goals and policies and sent back for review. A substantial number of counties have made important changes in their plans and implementation regulations in response to those Oregon-influenced goals and policies. For example, Hillsborough County (Tampa) submitted a plan for one dwelling unit per acre for thousands of acres of farm land. The plan was found not in compliance on this and other grounds, and the eventual plan for the land ranged from 5 to 40 acres per dwelling unit. A similar process occurred in Charlotte and a number of other counties.

Every state that adopts a growth management system promises the development community that the system will, through pro-active planning, lead to greater certainty and timeliness in the regulatory process. Sadly enough, *only* Oregon has been successful in delivering on that promise, which is doubtless part of the reason that corporate groups, and especially home builders, support Oregon's system. Through judicial, statutory and administrative reforms, Oregon has succeeded where other states have failed in simplifying and speeding up the time frame for permit approvals. Local governments have been mandated by the state to remove requirements that create uncertainty and delay, including a virtual ban on local moratoria on development. Oregon's success in this area continues to be a force as other states, with more recently adopted systems, struggle to achieve the same success. The implementation of Florida's law, now in its seventh year, has been a major disappointment to its supporters, and especially to the development community, in this area. With some exceptions, land development regulations have not been simplified and made more flexible, nor have permit time frames speeded up. Oregon's success in simplifying the permitting process continues to attract the attention of other states implementing or considering the adoption of state growth management systems. Florida's new Environmental Land Management Study Committee (ELMS III), established in late 1991, placed the matter high on its agenda for consideration.

There is extensive evidence that the watchdog role of 1000 Friends of Oregon has been a key element in sustaining and strengthening the implementation of Oregon's growth management system, and probably no other element of the Oregon system has been so closely watched and widely copied. Florida looked to Oregon as a model as it established its 1000 Friends of Florida in 1986. Florida's 1000 Friends organization has played a critical role in the early years of the implementation of the law, and has emerged in the 1992 session of the Florida legislature as the catalyst for mobilizing opposition to efforts to weaken the law. Other 1000 Friends groups have been or are in the process of being established in New Jersey (New Jersey Future), Maine, Washington, Massachusetts, California, and other states. Henry Richmond and his colleagues at 1000 Friends of Oregon have been instrumental in advising and encouraging the establishment of such groups, and Richmond has taken the lead in establishing the National Growth Management Leadership Project, a consortium of nineteen such groups to serve as a clearing house for ideas and action in the field of growth management.

Oregon's influence on Florida's growth management system has been substantial and ongoing. The 1992 session of the Florida legislature saw the first serious efforts to weaken the state growth management system, causing Governor Lawton Chiles to establish ELMS III in anticipation of political problems. As the state considered early corrections to its system, aiming for the 1993 legislative session, ELMS III looked at the Oregon record especially in the areas of combating urban sprawl, protecting farm and forest land, and affordable housing. The 1993 legislature responded by enacting several ELMS III recommendations that preserved the central concepts of the system but made it more user friendly.

At the same time, Oregon is looking closely at Florida's concurrency doctrine that calls for a "pay as you go" standard for meeting the infrastructure needs of development. Current efforts by LCDC, Portland's Metropolitan Service District, and others are exploring ways to use the concurrency doctrine to strengthen the link between land use and transportation. Sharing of experiences between the two states is frequent and ongoing, especially through their respective 1000 Friends organizations (DeGrove 1991).

Oregon's impacts on New Jersey

The New Jersey legislature in late 1985 passed a growth management statute, signed by the governor in early January 1986, that established a

State Planning Commission to be named by the governor to include citizens, local government representatives, and state agency heads. The commission was charged with developing a State Development and Redevelopment Plan that would guide in a general way growth patterns in New Jersey. The law was a response to twin forces that had evolved over the previous decade. One was a series of New Jersey supreme court decisions (Mt. Laurel One, Two, and Three) that imposed an affordable-housing obligation within a regional framework on New Jersey municipalities. In determining a particular municipality's fair share, the court used the state land plan prepared by the state Department of Community Affairs, but never formally adopted by the state. The state Supreme Court made it clear that it would continue to use this development plan unless and until the state adopted a new state plan. A desire to move the fair-share housing issue from the judicial to the legislative/administrative arena was a strong motivating factor in the adoption of the 1986 law. The other factor was the economic boom, especially in the construction of office park and other commercial space, that hit New Jersey in the 1980s (DeGrove 1987).

The Oregon experience in growth management influenced the New Jersey effort in several ways. A key meeting of several hundred legislators, local officials, environmentalists, state agency heads or representatives, planners, and others at Princeton University in 1986 heard major presentations about the Oregon experience in applying the consistency doctrine, combating urban sprawl, protecting farm and forest land, promoting affordable housing, and encouraging economic development through providing the development community with timeliness and certainty as important components of any growth management system. Henry Richmond of 1000 Friends of Oregon and John DeGrove have continued to give input through the State Planning Commission and the New Jersey version of the 1000 Friends organization, New Jersey Future. In addition, the State Planning Commission has established a peer review committee chaired by DeGrove, and including either Henry Richmond or Robert Liberty, former land use attorney with 1000 Friends of Oregon and now a private consultant. The committee has met annually to review progress in the development of the state plan. In these ways the successful experience of Oregon in drafting and implementing a growth management system has been a model held up for New Jersey to adapt to its own particular circumstances, with special reference to the functioning of UGBs in the protection of farm and forest land and in combating urban sprawl.

The development of the state Development and Redevelopment Plan has been slow and often painful in New Jersey, but the process eventually came into focus with the adoption by the State Planning Commission of the final version on July 12, 1992. From the first draft plan in April 1987 through two versions of a preliminary plan to an interim plan adopted in 1991, all of which have stirred considerable controversy in the search for a consensus, a number of central concepts that owe much to the Oregon experience have been incorporated in one way or another. Chief among these has been the development of a process that will result in the designation of certain areas for urban-scale development, and other areas to be kept rural for the protection of farm, forest, and environmentally sensitive areas. The final version of just how this will work is not entirely clear, but the concept has been sustained from draft to draft. A tier system ranging from the older central cities through to an environmentally sensitive tier has given way to "policy areas," but the central concept of separating rural and urban uses remains in the final plan. Much debate has centered around the degree to which the system will require municipal plans to be consistent with the state plan. A complex cross-acceptance process centering on New Jersey's twenty-one counties is aimed at a sustained negotiation process that supposedly will cause local plans to be consistent with the state plan and with each other. There is no direct requirement for consistency in the system, and the Florida and Oregon requirement in this area is still being discussed as a possibility if the cross-acceptance process fails to produce the degree of consistency needed for a responsible system.

Oregon's influence on Maine

Traditionally, Maine has been eager to promote economic development to provide jobs for its citizens. That tradition continued as Maine debated the adoption of a growth management law in the early 1980s, but there was a rising tide of concern about the negative impacts of growth, especially the relatively strong growth that took place in southern Maine through much of the 1980s. By 1986 many Mainers were alarmed at the effects of unmanaged growth on easy access to the shore, on lakes, rivers, and rural countryside, and on increasing traffic congestion in rapidly growing southern Maine. There was a widespread fear that Maine was at the mercy of these pressures, with no governance or policy mechanisms in place to manage growth properly. The influence of the print media in arousing the concern of their readers was substantial. A steady barrage of editorials and articles between 1986 and 1988 trumpeted the negative impacts of unmanaged growth. Headlines such as "Traffic Headache for Maine," "Fighting to Save Coastal Jewels," and "Taking Aim at the Land Speculators" were common in the two years leading up to the adoption of Maine's growth management act. One of the state's leading newspapers, the *Maine State Times*, prepared a series of articles describing the Oregon system and exploring how the system might be applicable to Maine. The articles were especially attentive to the use of UGBs to contain sprawl and conserve rural resources, both key emerging issues in Maine (Turkel 1988).

In August 1986, Maine's leading environmental agency, the Natural Resources Council of Maine (NRCM), published a comparison of growth management systems in Vermont and Oregon, and how those systems might be relevant to Maine. After examining the key goals in Oregon's law and assessing success or failure in implementing those goals, the study noted that two states, Florida and New Jersey, had "followed Oregon's lead" in enacting legislation to deal with the statewide problems of growth management. The report asserted that "experts in the Oregon program believe that it is exportable to other states," and that three essential elements were: 1) clear standards, procedures and definitions; 2) strong statewide political leadership; and 3) a broad-based public interest group such as 1000 Friends of Oregon to serve in a watchdog role in the implementation of the process. The council's executive director, Everett "Brownie" Carson, was himself familiar with the Oregon program and with the work of 1000 Friends of Oregon. Carson took the lead in making the NRCM a member of the National Growth Management Leadership Project, thereby sustaining Oregon's influence on planning in another state (O'Sullivan 1986).

The University of Southern Maine's Center for Urban Policy in 1986 and 1987 sponsored a number of conferences that brought in speakers from Florida, Oregon, and other states to help frame the issues more sharply for Maine. These meetings focused especially on the experience in Oregon with consistency, compact development, and the protection of natural systems. As one key actor in developing Maine's law put it, "The

NRCM Study convinced us that the Oregon model was the right one for Maine." Thus when Maine began giving serious consideration to the adoption of growth management legislation beginning in 1986, the Oregon experience was well known and viewed favorably by many supporters of the proposed legislation.

A report by Maine's state planning office in 1986 was the first official documentation of Maine's growth problems. The focus was on the cumulative negative impacts of unplanned growth, and it led to the naming of a working group in late 1986 that included most of the key concerned groups such as the Maine Municipal Association and the NRCM. Governor Joseph Brennan's state planning office drafted a proposed law that called for establishing state goals against which local plans would be reviewed and possibly approved, with a stronger role included for regional agencies. However, the proposal came in the last days of the Brennan administration, and when a Republican, John McKernan, won the governor's race, the report was shelved. The legislature took up the slack and authorized a Commission on Land Conservation and Economic Development made up of key legislators, a bipartisan group headed by the chair of the Joint Committee on Energy and Natural Resources. The Commission seemed united on the nature of the problems, but was divided on how to solve them, and especially on how strong the state role should be. Its hearings solicited input from all relevant stake holders, and received proposals in November 1987 from the Maine Municipal Association (MMA), the NRCM, and the Maine Real Estate Development Association (MREDA), with the new governor developing his own proposal a month later. The key disputed issues were: 1) whether local planning should be mandated; 2) whether local plans should be reviewed and approved by the state; and 3) what level of funding should be provided. Surprisingly, although the same pattern has repeated itself elsewhere, the call for the strongest state role came from the development group, MREDA. Fear of wildfire "no-growth" movements at the municipal level was probably the key reason for MREDA's position. In the three years preceding the adoption of the law, more than sixty municipalities in Maine adopted development moratoria of varying lengths to catch up with the need to better manage their growth. The MMA focused on the need for more state resources and technical assistance, but did not support mandated local planning or state review and approval of those plans.

Rep. Mike Michaud proved to be a skilled and persistent leader in winning approval of a strong recommendation that included mandatory local planning within the framework of a set of state goals, and review and approval of those plans at the state level. The report went to the legislature where the Joint Committee on Energy and Natural Resources, chaired by none other than Rep. Michaud, took it up. From January to April 1988, the committee held hearings and workshops, negotiated compromises, narrowly avoided a last-minute veto by the governor, and passed the bill into law as "an act to Promote Economic Development and Natural Resource Conservation." Michaud and other supporters of a bill with a strong state role and mandated local planning won in the end, with a compromise on the consistency issue.

Maine's law clearly reflects the influence of Oregon's growth management system, and certainly one reason was the considerable length of time Oregon's law has been in operation. The law contained a set of ten goals that set the framework for the system, comparable to Oregon's nineteen goals. The goals were broad in scope, including housing and economic development, as well as natural resource protection. A new state Office of Comprehensive Planning was established to give technical assistance to local governments and to review the mandated local plans "to ensure their consistency with the requirements of the act." The consistency requirement was not so strong, however, as in Florida or Oregon. Plans would be reviewed for consistency with state and regional goals and policies, but local governments were given the option of whether to have their plans "certified" as fully consistent with state goals and policies. The incentives to take that final step were extremely powerful, including the ability to levy impact fees and tap into a legal defense fund. It therefore seemed likely that most local governments would take the final step, and that appeared to be the case in the early implementation period. This consistency compromise brought the MMA into support of the bill, especially in view of the legislature's relatively generous funding of the mandated local plans. Again tracking the Oregon law, state agencies were brought into the system by requiring that the plans and programs of nine listed state agencies be made consistent with the state goals.

Concurrency and its focus on infrastructure were given status in the law by calling for "an efficient use of public facilities" to accommodate anticipated growth and economic development. The bill called for compact urban development patterns with related protection of rural areas from unplanned urban development, with "orderly growth and development in appropriate areas of each community" and "making use of public services and preventing development sprawl." The compact development requirement was not so clear and strong as in Oregon, but it was there. No absolute concurrency requirement was included, as in Florida, but the thrust of the goals and policies was strongly in that direction.

As in Oregon, economic development was given goal status, as were water resource protection, critical natural areas, and agricultural and forestry resources. The housing goal called for 10 per cent of all new housing in Maine to be affordable for low-income households and a bond issue of $35 million authorized in the same year (1988) earmarked money for both housing and the acquisition of public lands to protect natural resources. In short, Maine's law made it a leader in new state growth programs of the 1980s, and while the system was influenced by many groups and individuals, the evidence is clear that the major external influence on the program was Oregon (DeGrove 1991).

Oregon's influence on Washington state

The same negative impacts of unmanaged growth that caused Oregon to adopt Senate Bill 100 in 1973 were present in Washington at the same time. One might have expected Washington to put a growth management system in place in the mid-1970s instead of fifteen years later. In fact Washington began the serious consideration of such action in the early 1960s when a citizens group, reporting to a joint committee of the legislature on urban area growth, called for a new metropolitan governance system to manage the heavy growth pressures the state was experiencing. The state was a leader in adopting a Shoreline Management Act, but despite efforts by Governor Dan Evans a comprehensive state land use law failed to pass the legislature in the early 1970s, at about the same time Oregon was evolving its program. A recession in the aerospace industry in the late 1960s, and a timber recession in the late 1970s and early 1980s (shared by Oregon), put comprehensive growth strategies very far back on the back burner.

By the mid-1980s, with the economy booming and increasing fear of a population invasion from California and elsewhere, a push for a state growth management system reasserted itself. The negative impacts of substantial growth, especially in the Puget Sound (Seattle) region, were the subject of a major leadership conference sponsored by the Seattle Chamber of Commerce, and a series of articles on the same subject commissioned by the *Seattle Times* from nationally syndicated columnist Neal Peirce (Peirce and Johnson 1989) heightened interest in the issue. Governor Booth Gardner supported action to deal with the problem, and a legislative champion emerged in Joe King, speaker of the State House of Representatives. In a rapid series of events a Growth Strategies Commission was called for by the legislature and named by the governor in late 1989 and the 1990 legislature passed phase one of a Washington growth management system even before the Growth Strategies Commission completed its work. In September the commission produced a strong report calling for the adoption of phase two to put implementation teeth in the goals and policies put in place by the 1990 legislature; and the 1991 legislative session did adopt phase two of the system.

Taken as a whole, Washington's growth management system is one of the most powerful in the nation, at least for the 26 of Washington's 39 counties and the cities within those counties to which it fully applies. The system draws heavily on the experience of Florida and Oregon. From Florida it drew the concept of concurrency, from Oregon the concepts of urban growth boundaries, the protection of farm and forest lands, and economic development. Experts on those programs were featured at growth management conferences and as speakers at meetings of the Growth Strategies Commission. Furthermore, the Washington Environmental Council, the state's leading environmental group, has close ties to 1000 Friends of Oregon. The council played a strong role in developing a growth management proposal of its own, much of which was incorporated into speaker King's proposed legislation.

When we examine the content of the 1990 Growth Management Act (HB 2929) and the 1991 legislation (HB 1025) in tandem, we see a system that is one of the most demanding in the nation, and draws more from the Oregon experience than any other one source.

The concept of consistency did not make it into HB 2929 in any clear fashion, and there was no state or regional oversight to assure that local/regional or state agency plans would be consistent with the thirteen goals contained in the law. Internal consistency within each plan was required; all local plan elements, which had to include land use, housing, capital facilities, utilities, rural, and transportation elements, were required to be consistent with the future land use map. In HB 1025 the consistency requirement was expanded to include local, regional, and state plans, and a review and consistency determination process was put in place. The overall thrust of HB 1025 was to provide the state oversight and enforcement provisions that had been missing from HB 2929. Three Growth Planning Hearings Boards were appointed by the governor for each of the three major regions of Washington. These boards were to hear challenges to state agency, county, or city plans as not being in compliance with the requirements of the act. If a plan is not brought into compliance, the governor has a wide range of sanctions involving the withholding of funds.

HB 2929, borrowing from Florida, contained a concurrency requirement for transportation facilities. Levels of service for roads were required to be set, including "regionally coordinated levels of service," and backlogs and future transportation needs had to be identified for a ten-year period. Local governments were required to "adopt and enforce ordinances which prohibit development approval if the development causes the level of service on a transportation facility to decline below the standards" adopted in the comprehensive plan. The bill also contained as one of its goals the expansion of the concurrency requirement to include other key facilities such as sewers and water supply. This goal remains as one of the guidelines in HB 1025.

The third major concept included in HB 2929 in 1990 was a requirement that local governments establish Urban Growth Areas (UGAs). The concept was borrowed directly from Oregon, and it is a strong one. Counties were required to work with cities to establish the growth line, but if agreement could not be reached it was up to the county to take action, with cities allowed to appeal to the Department of Community Affairs, which would mediate the dispute. The criteria for setting the line were much the same as in Oregon: existing cities and urbanized or urbanizing areas were to be included, with densities high enough to meet twenty-year population projections. UGAs were to be reviewed every ten years and, if needed, densities increased to accommodate population growth. The 1991 law strengthened the UGA requirement by mandating countywide planning policies that would include, through a collaborative process with the cities, setting the UGAs. It is important to note that the countywide planning policy also has to include provisions encouraging compact development within UGAs, affordable housing, economic development, transportation facilities, and a provision for joint city-county planning within UGAs. A multi-county regional planning policy was required for counties of at least 450,000 population, namely the three Puget Sound counties of King, Pierce, and Snohomish. This tri-county approach is similar to the three-county Portland Metro area in Oregon.

In a departure from the Oregon urban growth boundary approach, HB 1025 addressed the issue of what to allow outside the UGA boundary. Oregon's SB 100 allows nothing in the way of substantial new urban development. Florida law and rules on the issue are still evolving. HB 1025 provides for planned free-standing communities outside UGAs under strict conditions to prevent erosion of rural and environmental values. Two categories, fully contained communities and master planned resorts, are included. Compact development, a transit orientation, buffers, and land use regulations to assure the maintenance of rural densities around such development and the mitigation of damage to farm and forest land are spelled out in some detail.

The issue of economic development for eastern Washington's poorer counties has always been high on the growth agenda in Washington state, and HB 1025, reflecting at least some of the recommendations of the Growth Strategies Commission, contained a package of policies and requirements to encourage economic development. Natural resource protection, including especially the protection of agricultural, forest, and mineral lands, and the identification of natural resources of statewide significance, were included in the legislation.

Washington state adopted a growth management system almost twenty years after Oregon acted, and it drew on the Oregon experience heavily, though certainly not exclusively, in shaping its own system. Once again, Oregon's relatively long record in implementing a comprehensive growth strategy provided a fertile field for yet another state to borrow from the Oregon experience.

Other states influenced by the Oregon system

The fact that Oregon has had such an impact on a large number of state growth management systems that have developed in the 1980s and continue into the 1990s is logical in that the Oregon program is comprehensive, has a relatively long implementation record, and has some impressive accomplishments to recommend it. On the other hand, the Oregon program for a long time after its adoption was viewed by many uninformed people and by others who either should have known or did know better as 1) a no-growth effort that aimed somehow to stop everybody at the state line unless they only planned to visit; 2) driving the cost of housing through the roof because of land limitations imposed by urban growth boundaries; 3) subversive of private property rights; 4) stifling of economic development; and 5) other negatives too numerous to list. Only in the last eight or ten years have we begun to see careful descriptive and analytical research on how the system actually works. Since that has roughly coincided with the emergence of new state growth strategies in the latter half of the 1980s and now into the 1990s, the Oregon influence on other state programs has indeed been substantial.

Vermont and Rhode Island both adopted, along with Maine, a state growth management law in 1988. Both of these states' laws were influenced by the Oregon program in much the same way that Maine was. First, the key actors in developing legislation in the three states have been in more or less close communication with each other. The Vermont Natural Resource Council interacts closely with the NRCM, and with comparable environmental groups in Rhode Island. The three laws have much in common, including a set of goals that frame the system, and to a greater (Rhode Island) or lesser (Vermont) degree, those goals must be reflected in regional (Vermont only) and local plans through the doctrine of consistency. Concurrency, compact urban development/anti-urban sprawl strategies, affordable-housing requirements, the protection of natural resources, and economic development are reflected in the goals and implementing strategies. Reasonably generous funding, at least until the recession year of 1992, was a feature in all three programs for both local and regional governments in crafting and implementing their plans. Adding to the communication within the three New England states were a number of seminars before and after the adoption of the laws in which the "Oregon message" was well represented.

Rhode Island's law is arguably the most powerful state growth management system yet adopted, in that the mandate for state agencies to develop plans and programs is very strong. In addition, state agencies are prohibited from placing a project that is inconsistent with a local plan in the boundary of a town with a consistent plan without an extensive public hearing process. The consistency requirement for local plans is clear and firm, and if a town fails to develop such a plan, the task is undertaken by the state. Of course there are only 39 municipalities in Rhode Island, so such an approach is feasible there. Affordable housing, economic development, compact development strategies including redevelopment, and the protection of natural resources are all incorporated in the goals and implementing regulations, which are also mandatory (DeGrove 1991).

Vermont is a more complex case. The report of the Growth Strategies Commission (the Costle Commission) to Governor Madeline Kunin called for a strong state law that mandated local plans and land development regulations consistent with the twelve goals developed by the commission after extensive hearings across the state. In the give and take of legislative consideration of the commission's (and governor's) recommendations, the goals were expanded to 32, and mandatory local planning was deleted. However, if a local government chose to plan, its plans and regulations had to be consistent with the state goals and certain regional policies, and with each other. The key plan review function was given to strengthened regional agencies. The implementation of the law, Act 200, at first smooth enough, ran into trouble in the 1990 legislative session and the law narrowly escaped repeal. It did survive but with deadlines extended and consistency review at least temporarily softened. Similar efforts to undermine the law seem to have failed in the 1992 legislative session, and the plan development and review process is moving forward. There are substantial incentives to encourage towns to develop plans, such as access to impact fees and other financial support for both planning and infrastructure. Funding, originally adequate, has been cut back in the face of the recession, which will delay the process further. Still, the program is surviving if not thriving, and except for the mandate to plan it contains the essential elements, the concepts and principles, common to the state growth strategies of the 1980s and 1990s. The Vermont state growth strategy has the strongest affirmative affordable-housing component of any of the programs, developed through a Housing and Conservation Trust Fund established in 1987, and has been generously funded since, including an added $11.4 million by the 1992 legislature, though it was forced to cut deeply in almost every other area. The law has a strong economic development focus, and that too received added funding by the 1992 legislature.

In Conclusion

Oregon adopted SB 100, the statutory base for its growth management system, almost twenty years ago. Often embroiled in controversy, and frequently criticized by friend and foe alike, the system nonetheless has survived and in many ways thrived. Its accomplishments as well as its weaknesses are detailed elsewhere in this volume. The fact remains that as new states turn to the task of crafting their own state growth strategy, they always look to Oregon for ideas, and for some understanding of what works and what does not. Perhaps most important of all, the recent establishment of the National Growth Management Leadership Project (NGMLP) means that the cross-fertilization of ideas with regard to state growth management systems will be stronger than ever, and Oregon will be at the center of this emerging network.

The NGMLP has put in place an information exchange system that has the effect of further extending the Oregon experience across the nation. Nineteen states are now members, regular meetings are held, and a quarterly newsletter titled *Developments* has featured a number of articles on the Oregon experience, beginning with an article in its first issue titled "Oregon: 15 Years of Land Use Planning" (Kasowski 1990) which noted that "in tracing a course on the map to new growth management policies, the state of Oregon is often used as a prominent reference point." Other articles have featured Oregon's experience in protecting farm land, providing affordable housing, combating urban sprawl, and others. The self-stated purpose of the NGMLP "is to provide a quantum leap ahead in developing new state and regional policies to better manage growth." Developed under the leadership of 1000 Friends of Oregon and funded largely by foundation grants, the existence of NGMLP means that the Oregon influence on growth management systems in the 1990s will continue, and indeed that influence is clear in the content of a state growth management system proposed by the governor and adopted by the 1992 session of the Maryland legislature. It seems clear that through the NGMLP and in other ways, the Oregon experience in growth management will influence in important ways new growth management programs being considered by states in the 1990s.

References

DeGrove, John M. 1984. *Land Growth and Politics.* Chicago: American Planning Association Planners Press.

———. 1987. "Balanced Growth in Florida: A Challenge for Local, Regional and State Governments." *New Jersey Bell Journal* 10, 3: 38-50.

———. 1991a. "Managing Growth in Other States: A National Perspective." Paper presented at Conference on Balanced Growth: Promoting the Economy/Protecting the Environment. College Park: University of Maryland, November 22, 1991.

———. 1991b. "Regional Agencies as Partners in State Growth Management Systems." Paper presented at Planning Transatlantic: Global Changes and Local Problems, Joint ACSP and AESOP International Congress, Oxford, U.K., July 8-12, 1991.

Florida Department of Community Affairs. 1989. Technical Memo. Volume 4 Number 4.

Kasowski, Kevin. 1990. "Oregon: 15 Years of Land Use Planning." *Developments* [National Growth Management Leadership Project Newsletter] 1, 1: 6-8.

Mount Laurel One—*Southern Burlington County NAACP v. Township of Mount Laurel*, 67 N.J. 151, 336 A.2d 713, Appeal Dismissed and Cert. Denied, 423 U.S. 808. (1975).

Mount Laurel Two—*Southern Burlington County NAACP v. Township of Mount Laurel*, 92 N.J. 158, 456 A.2d 390(1983).

Mount Laurel Three—*Hills Development Co. v. Township of Bernards*, 103 N.J. 1, 150 A.2d 621 (1986).

Peirce, Neal, and Curtis W. Johnson. 1989. "Congestion and Sprawl; Sprawl Stalks a Stunning Natural Treasure; The Bitter Harvest; No One's in Charge; Politics of Postponement; A Critical Need for Learning and Training . . . and Caring; A Way to Wed Conservation and Development." Seattle *Times*. October 1-6, 8.

O'Sullivan, Karen Pettetier. 1986. "An Examination of Land Use Legislation in Vermont and Oregon, and a Comparison with Maine." Presented for the Natural Resources Council of Maine, August 1986.

Turkel, Tux. 1988. "Opponents hammer out accord on controlling growth in Maine." *Maine Sunday Telegram*. April 17.

CHAPTER 12
Managing "the Land Between": A Rural Development Paradigm

Robert C. Einsweiler & Deborah A. Howe

I. The Problem Situation

Between the edge of suburban development and operating agricultural lands, one generally finds a belt of speculative land holdings which we might call, paraphrasing Charles Little, the "land between" (Smithsonian 1979). It is a fitting name. Agricultural lands as extensive uses and urban lands as intensive ones cannot coexist in the same competitive land market; the urban uses can outbid the agricultural ones.[1] Put another way, the economic return from an acre of urban land is so much greater than from an acre of agricultural land that potential users for urban purposes can pay more for the land. The "land between" occupies the space between higher urban and lower agricultural land values—too high priced to farm and not yet needed for subdivision. Agricultural investment is discouraged and "buckshot urbanization" emerges (Little and Fletcher 1981). If land values responded only to actual demand, this land in waiting would be priced the same as surrounding agricultural land. The higher values are expectation or speculation values that are betting on future urbanization.

The Oregon system intervenes in this real estate market by establishing urban growth boundaries (UGBs) around incorporated cities. Development is to be contained within the boundaries and lands outside the UGBs are designated for resource use. The line between urban development and resource use is expected to produce a stairstep in the otherwise declining slope of land values (from the city center outward), to match the investment abilities of the two competing markets. Urban-level speculative land values are then limited to areas within the UGB. To achieve this outcome with the UGB as the primary policy instrument, the boundary must remain stable, giving investors confidence that the differential values will prevail.[2] Research suggests this has occurred, with farm land values reflecting agricultural rather than urban development potential (Knaap and Nelson 1992). Thus, in theory, Little's notion of "the land between" disappears in the Oregon context.

But reality is different. Existing development patterns outside of UGBs created a problem in this otherwise simple, two-part system. This development did not fit neatly into either category. As a result, counties were allowed to take exceptions to the resource goals when parcel size and land use patterns precluded farming and forestry. These "exception" areas include the low-density scattered development and the hundreds of unincorporated communities that serve as the foci of rural life. Scattered throughout the state, the rural communities range in size from 2 acres to a square mile or more with fifty to five thousand residents. Exception areas are theoretically available for rural development. They are in essence a legitimized form of "land between."

To gain some measure of the problem, of the state's private lands, 8 percent are urban and 88 percent are farm and forest. The exception areas make up the remaining 4 percent, equivalent to one-half of the urban category. The state, however, has not developed a sufficient policy framework for guiding these areas. None of the nineteen statewide goals specifically address rural settlements (although Goal 11 requires rural public facilities planning). As exceptions to the state's agricultural and forestry goals, these settlements are the equivalent of nonconforming uses in standard zoning; they are not uses in their own right. This negative orientation toward rural development conveys a sense that the rural lifestyle is not a valued part of Oregon society and thus of the planning system. The counties are restricted in trying to plan for rural development, since the Oregon system constrains the provision of urban-level services outside of UGBs and does not allow the expansion of exception areas. This limits alternatives for responding to changes in the economy such as increasing tourism and significant declines in Oregon's forest industry.

The Oregon Supreme Court spoke to this issue in the 1986 Curry County case. It ruled that taking exceptions to the agricultural and forestry goals does not alleviate the need to take exceptions to the urbanization goal when providing for urban-level densities outside UGBs. Put another way, the counties must not only make the case that these lands could not or should not be resource lands, but also why they could not or should not be urban. The court asked the state for a clear policy on rural land.

In 1991 Keith Cubic, Douglas County planning director, submitted a petition to the Land Conservation and Development Commission (LCDC) requesting rule making to address the problems raised in the Curry County case. Twelve counties wrote letters of support. In a "friendly" rejection of the petition, the commission encouraged Cubic to proceed with developing a consensus on rural policy. LCDC agreed to put the development of a rule for rural communities in their work program; this effort began in mid-1992, but was put on hold during the 1993 legislative session.

The task is not easy. The simplest solution would be to designate rural development as a third major category. But the two-part division of the markets via regulation is essential if no acquisition of property rights is to occur, or no subsidy is to be given to offset speculative land prices in resource areas, or no tax is employed to discourage speculation. For this regulation-dominant approach to work, it must be rigid. This rigidity, however, makes it difficult to take into account locally specific variations in how resources are or can be used and in the nature of urban areas. Furthermore, the eventual need to move the line as growth occurs could reduce its ability to shape the market into two price levels, since speculation would again become a factor.

This chapter represents an effort to enter this issue by suggesting an alternative way of thinking about urban development, rural development, and resource lands, and the means of managing them. We will focus in part II on the emerging urban and natural resource land use patterns and in part III on why governments intervene in land markets and how they intervene. In part IV we will apply these frameworks to Oregon and suggest ways to think about alternatives for this land between.

II. The Emerging Land Use Pattern

The urban pattern

As late as the 1950s, urban places in the United States tended to have hard edges. They had centers called central business districts (CBDs). At the center of the CBD was the 100 percent corner, a peak of pedestrian traffic that often occurred where two department stores shared an intersection. Commuters lived in suburbs, but worked in the central city. There was generally one worker in the nuclear-family household who was the husband and father. One auto handled a family's personal transportation. Central place theory was alive and well; the organization of urban places could be explained by the hierarchy of trade (Christaller 1933, Alonso 1964, Muth 1969). The key point is that urban places were discrete, with a highly concentrated pattern of employment. There was limited commuting into and out of these places. The state-level urban pattern could be seen as dots in space.

The automobile began reshaping urban regions in the 1950s and continued to do so until accessibility became fairly uniform. Global economic restructuring started transforming nonresidential uses within urban regions in the 1970s and accelerated the pace in the 1980s (Hart and Mayer 1991, Ladd and Wheaton 1991). Trade centralized up the hierarchy (Anding et al. 1990). CBDs in major centers were linked to the global information network and were as much a part of that pattern as of their own physical place (Hall 1991). Manufacturing decentralized to the suburbs, to rural communities, or overseas (Butler 1991). Offices and many services emerged in the suburbs. The suburban centers of the metropolitan plans of the 1960s were observed by many authors, but Joel Garreau's book naming them "Edge Cities" made them real in daily conversation (1991). Polynucleation that began with enclosed suburban malls as early as the late 1950s became the new urban pattern across the globe. Even so, two-thirds of the employment in at least one major urban region was not in centers, but dispersed across the urban area (Giuliano and Small 1991). Two-worker couples with multiple autos crowded the streets and highways. The size of housing lots continued to grow and overall densities of new residential areas declined.

Within these regions of urban influence are rural communities. Nationwide, 46 percent of all rural population lives in metropolitan areas as defined by the census (Schwartz 1990). Commuting pulls out the boundaries of metropolitan areas and in so doing transforms the nature of rural communities. Weekend tourism and second-home development also blur the distinction between urban and rural. Within all this change, however, physical place seemed to still have some constants. In at least one setting, the pattern of locations and relative community size remained constant for a century, although individual places grew or declined (Hart and Mayer 1989). Urban places still can be viewed from the air as physical places having an edge, although a more fragmented one. But there is a difference. What has been added are commuting and other travel bands connecting multiple residential communities to multiple work places, some of which are now located in rural areas. In *function* we have linked, multi-centered urban places, although in *form* they may appear to be physically separate.

Moving further into the notion of linked centers in rural regions are "cluster communities" which are not apparent by reviewing census data in normal form (Center for the New West 1992). These are networked communities, nearby but not abutting communities in rural regions. They join together for economic development and other purposes without creating a new overarching governmental structure. They are true horizontal, not hierarchical, clusters typical of all networks. So far as is known, this phenomenon began almost twenty-five years ago in Alberta, Canada, and spread to Saskatchewan, Minnesota, Michigan, and Iowa. They have never appeared as entities in standard secondary-data sources since they officially are not places or minor civil divisions. They may therefore exist in other locations as well. Known clusters encompass from two to dozens of communities (the large clusters are in Canada), within or across county boundaries. In 1990, Iowa made them eligible for state funds and some federal pass-through money.

The United States and Canada, and perhaps the rest of the world, have moved from the implicit notion of discrete, autonomous places to a network of communities connected by commuting or other linkages. Christaller's observed world of central places did not rest on the view that the adjacent population was tied to this single center. The centrality was a result of the available modes of transportation and structure and levels of income in that day. This reality would have been a perfect fit to Oregon's UGBs since there would have been little demand for significant numbers of people to live in the natural resource areas.

But today's stress on the two-part urban-resource concept is different. There are land-value pressures from those who wish to live immediately outside the UGB and from the longer distance commuting that undoubtedly occurs between exurbanites living in exception zones and various urban places (Davis 1990). This commuting also places pressure on the functioning of the resource lands. It interferes with use of roads and acceptance of resource management practices (odor, spraying, and the like). It probably extends the overlay of speculative land values even further from the UGB. This combination places greater burdens on the expectation that separate urban-level and resource-level markets can be maintained. Last, but not least, the Oregon system's simple view that equates urban with only incorporated places is out of touch with reality.

The natural resource pattern

In the Oregon system, natural resource categories are basically economic uses—farming and forestry. Another way to consider the use of natural resources would be along a spectrum from 1) an economic surface for development (the traditional property rights view of urban land) through 2) production lands (farms, forests, fisheries, mineral lands) and 3) natural process lands (aquifer recharge zones, wetlands, and the like) to 4) ecological communities. The first two are economic uses for which land is exchanged in markets; the latter two command no market prices. Yet all four are Mother Nature's contributions and are needed and valued by society at large.

As one moves from category 1 to 4, our understanding of the natural functions occurring on the land decreases. Therefore, if we wish to intervene and convince a court of our need to do so, our ability decreases as we approach category 4. Put another way, we fully understand the calculus of the economic uses or investments in categories 1 and 2. We also have quite strong understanding of resource attributes supporting foundations, septic tank drainage, and the like in category 1, the natural processes of raising plants, growing trees, and the like in category 2, and the damages to nature that occur in this realm—erosion, flooding, subsidence, and wildfire in forests. However, we have less understanding of how much interference with the functions of resource extraction can be tolerated such as commuter traffic on roads used by tractors.

When we move to category 3, the processes of nature are largely understood through physical sciences—aquifer recharge, absorptive capacity of a roiling river, the capacity of the atmosphere to transport and absorb—our models grow weaker. Global warming prognostication is one example of weak modeling in a very complex atmosphere. Finally, in group 4 situations, we must rely on the life sciences to understand wildlife habitat and population ranges and dynamics. Here our grasp is weaker still. This weakened understanding of the natural realm must confront the fairly strong understanding of diminution of value from the economic realm. Thus we do not know with certainty if and how we can make use of old-growth timber resources without destroying the habitat of the northern spotted owl. We have a clearer idea of the toll in unemployment, lost local government revenues, and lost revenues to firms if timber harvest is not allowed to continue.[3]

This four-part spectrum reflects the stages of intervention of environmental considerations via land use regulations into land markets. It also is indicative of the increasing awareness of the citizenry at large about environmental damage and it parallels environmental pollution legislation (constraining the side effects of development on the environment). The same spectrum is found to a great extent in the existing Oregon system. The UGBs by and large work like the first category. Resource lands are category 2. These uses are economic activities like manufacturing, but they make more extensive use of the land and they use its resource value, not just its surface value. Other statewide goals relate to the third and fourth categories, but they are often not treated spatially in local plans. They definitely are not treated as uses in zoning ordinances, as they are not considered to be legitimate economic uses.

Given that use decisions by individuals tend to be based on economics, category 1, the human settlements category, can produce negative impacts on the remaining three categories. In similar manner, farms and forests can produce negative impacts on the remaining two categories (and on the urban category if it is too close) depending on the degree of care taken in their operation.[4] But as human settlements become smaller, contain less people and less nonresidential activity, produce less traffic and sewage, and the like, the degree of care necessary to protect resource lands diminishes. The technology used in resource management also has a bearing on the potential level of conflicts. Timber harvesting that depends largely on draft animals for selective cutting has a very different set of impacts on resources and side effects on neighbors than high-technology machine cutting.

III. Framework for Managing Land Use

Why governments intervene in land markets

Governments intervene in land and real estate markets to achieve public objectives or patterns of development that private developers do not produce on their own. They also intervene to reduce or eliminate the negative effect of private decisions on others in the community.

There are five main reasons that real estate markets do not produce satisfactory results without government intervention. First, the transactions do not involve all who are affected. Side effects of traffic, noise, glare, the character of a neighborhood, and loss of views fall on neighbors. Fiscal impacts from providing supporting infrastructure or services fall on the municipality. All this can change over time. Owners can alter how they use land, producing effects that were not anticipated at the time of the transaction. By not including these side effects as a development cost, a project is effectively given an unearned subsidy.

Second, the incremental nature of these transactions, summed over time, produces less order and less effective functioning of the community than residents desire. An imbalance between employment and the provision of housing is one example of inefficiencies in metropolitan land use patterns.

Third, real estate transactions are basically economic ones. They use a traditional economic calculus that equates costs and benefits to income and expenses. This ignores effects on assets and liabilities. Resources sold off generate income. The income increase is captured, but the loss in resource assets is ignored (Repetto et al. 1989). Further, standard economic analysis treats natural resources—air, water, the environment—as free goods and as "sinks" for pollutants. And it ignores social concerns that are not part of this development production/consumption process.

Fourth, the timing of development decisions, which historically has been left to the private sector, can create problems in the provision of adequate public services.

Fifth, because land is finite and fixed in location, and development decisions are so durable, the normal compensating mechanisms and theories of markets in goods and services quickly consumed do not apply or apply only partially. Land has an aspect of monopoly based on location that is not a characteristic of commodities.

How governments intervene in land markets

Given these market imperfections, public intervention is warranted. But it also is circumscribed. Under the constitution, individuals have development rights in land to the extent not constrained by law. In light of these rights, the purpose of government intervention is to alter how private and public investments are made. The process of development and its management can be described as an interaction between the public and private sectors.[5] The private side of the relationship, unconstrained by government, responds to consumer concerns. It does not attempt to achieve the larger social concerns of the community unless that is necessary to its prime purpose. The private sector draws on the natural environment to enhance its product and uses it as a place to dispose of wastes unless constrained by government or the needs of the project.

The government represents its own interests, principally provision of infrastructure and other necessary services. It also speaks for the society at large or enables this expression through participation in decision making. And it speaks (at least in theory) for the environment, which has no voice in economic markets. Government, in effect, brings others into the transaction.

The government relates to the private sector through legal authority over the private property owner or through financial mechanisms that directly affect the profitability of the project. The traditional legal authorities employed are eminent domain and acquisition, and the police power or regulation (the primary emphasis of the Oregon system). The financial mechanisms derive from the taxing and spending powers and also from the police power. These include special assessments, capital improvements, impact fees, other taxes and charges, and favorable-rate loans. As development projects become more complex, a development agreement is used to set forth the mutually agreed-upon public and private sector roles. This two-party agreement, a creature of contract law, frequently allows no public participation. As a consequence, citizens often intervene via initiative and referendum, a procedure that squeezes both the developer and the government out of the process.

When markets are strong, as in rapid growth areas, the regulatory devices or restraining methods are predominant. When markets are weak, as in redevelopment or economic development areas, financial mechanisms are used by government to share risk, reduce costs, or otherwise support the developer to go beyond the market in achieving public objectives. More recently, a mix of regulatory and fiscal devices have been employed to relieve the burden on a single technique.

In the more common situation, rural and resource lands are seen as eventual urban lands. Only the market determines when and where the transitional boundary, the margin of cultivation, occurs. Oregon has taken an unusual tack in declaring resource lands to be valuable assets that are not to be treated as future urban sites. Having done so, the state placed itself squarely in the conflict between the differential strength of the land markets for these two categories and imposed urban-style land use regulations on rural lands.

Alternatives for controlling development

There is great variability in the way in which growth management systems tackle the challenge of land use guidance. Three considerations of system design have a bearing on this discussion.

The first is a way of thinking about intervention, the systems strategy. This can span the spectrum from a design-like framework with spatial mapping of discrete areas (a focus on the desired result) to a market framework in which interventions are geared toward influencing land values and market choices (a focus on the mechanisms that will produce the result). The Oregon system leans toward a focus on the desired result in its emphasis on drawing boundaries around resource areas and urban developments. We will suggest the benefits of also using the market view.

The second is through use categories. As suggested above, Oregon defines two main uses, but the catch-all of exception lands as a third category weakens the clarity of the two-part system. We will suggest an alternative two-part framing.

Finally, there are techniques of intervention including specifying use through exercise of authority, allowing greater market decision making through performance regulation, and using fiscal tools to regulate. Oregon uses all of these, but the dominant techniques are traditional specification of use. We will explore the potential for expanding the use of other techniques.

With this context, we will now discuss both historical options and some more recent approaches for guiding land use development. Growth management systems generally make use of three different strategies, separately or in combination, to alleviate the side effects of development: controlling geographic space for specified uses, managing support systems or infrastructure necessary for development, and focusing on compatibility through performance or mitigation requirements.

Systems that control geography, that use spatial bounding, and focus on location. These systems have a design-like framework. They implicitly accept a single market in which urban use is free to outbid natural resource use. Intervention techniques are historically oriented to specifying the end result, as in zoning. In specifying use by location, community judgment is substituted for the operation of the market on these variables. Side effects are controlled by physical separation of uses and urban functioning by the ordering of uses. The task of specifying use by location has two significant hurdles. It must outguess what the demand for uses in the marketplace will be. And it must create a use category that is reasonable for the owner, the supply side of the equation.

Another way to think about the difficulties of traditional zoning is that zoning is a system of specifying end-state conditions, although the primary concern should be with managing transitions. These include spatial transitions between a use and its neighbors, and transitions on a site over time as in rural to urban conversion or redevelopment. Put another way, traditional zoning manages development as though it is static rather than dynamic. That is why the UGB technique is a difficult one to adapt to change.

Conceptually, the side effects of one use on another are controlled by physical separation of uses. However, while separation may have achieved some benefits, it also created costs—such as traffic—that a more mixed development would not have generated. Further, spatial separation has costs that impose limits. For example, sideyard requirements are costly; when sideyards are squeezed they can only contain limited side effects.

At the boundary between commercial and residential zoning one may find residential deterioration. This could stem from the negative effects of traffic, noise, and glare from the commercial property, so that the site becomes less desirable for housing and loses value if there is no potential for commercial use. Or it could be the result of a rise in value based on expectation of commercial use, making residential use unprofitable, so that the owner disinvests and waits for the appropriate time to capture the speculative gains.

A further difficulty is that the side effects of a specified use can change over time even though the use remains within its allotted category. At the time zoning was invented, the grocery business had walk-in customers and home delivery. Then came supermarkets with auto traffic followed by the hypermarket at one end of the spectrum, the supermarket in the middle, and the neighborhood convenience store. Each version of the grocery store has different impacts; yet each could fit within the same zone with a large enough site. To accommodate all these variations in use, zones that were large and limited in number in Germany in the 1890s have been subdivided many times over into a vast number of categories since zoning was imported to the United States (Logan 1976). In the 1980s, global economic restructuring has resulted in urban restructuring which in turn is affecting how land is and will be used. Outguessing the market with an increasingly particularized system has grown ever more difficult.

When uses change over time through rezoning, this creates a windfall for the property owner that provides a great incentive to manipulate the system. This is the dilemma Oregon faces at boundaries between urban and natural resource areas.

To overcome these various weaknesses, a variety of additions to standard spatial bounding systems have been invented—conditional use, design standards, and performance criteria. These are discussed below.

Traditional zoning is the dominant form of land use regulation in Oregon cities and counties. The Oregon system itself is in essence zoning writ large: uses are specified by location. Its categories are large and general. And one of its uses—rural development—is not really a category, but rather a nonconforming use existing as an exception to agricultural, forestry, and urbanization goals.

A more recent use of spatial bounding is the concept of policy areas or tiers. The idea was first used in the Twin Cities metropolitan system in the 1970s, refined and applied at the city scale in San Diego, and further extended in the New Jersey state plan. This approach to spatial bounding identifies otherwise similar areas where a single management technique will be employed differently to achieve varying policy objectives. The most common purposes are to differentiate based on environment or for infrastructure investment.

Systems that control infrastructure and use it to grow from nodes or existing development. This approach is in the middle of the spectrum and is usually selected when fiscal stress is the key growth management issue. One of the earliest local growth management systems of this type was in the town of Ramapo, New York (Einsweiler et al. 1975, Gleeson et al. 1976, Godschalk et al. 1979); it addressed only the area surrounding incorporated villages. These villages were to be the locus of high-density residential development and commercial activity. The prime mode for managing growth was the adequate facilities ordinance. By requiring adequate facilities as a condition of development, and by controlling the provision of facilities through the capital improvement program, the town determined the timing and sequencing of development.

The state of New Jersey uses node-oriented concepts in addition to the tiers in its state plan by creating a Regional Design System (New Jersey State Planning Commission 1988). The scheme has three related components: 1) a hierarchy of central places with varying functions—city, corridor center, town, village, and hamlet; 2) five types of development—redevelopment, in-fill, fringe, new centers, and rural; and 3) policies about services in these areas, including priority, type, and degree of care in installation. The Regional Design System reinforces the idea that development should be concentrated in and around existing communities. It adds to this the concept that communities have unique socioeconomic functions within a system of communities, so that connections to other places are important. Planning for each community needs to consider its context among other communities so that public investment and policy can reflect and reinforce these differences. Although the scheme does not explicitly endorse the idea of networked communities, it recognizes the relationships by specifying that transportation connects them into regional networks. As in Oregon, there is to be a clear delineation between urban and rural lands.

Because New Jersey is the most densely populated state in the nation, the system may well be more complex than needed in Oregon, but the underlying ideas may have applicability. Unlike New Jersey, the Oregon system currently classifies unincorporated places as rural and incorporated places as urban regardless of their size or the function they perform in the economy. In this scheme, Burns, a rural center in eastern Oregon with a 1990 population of 2,913 is treated by the state planning system in the same manner as Salem, the state's capital, which has 107,786 residents and is part of a larger metropolitan area containing 278,024 people (Center for Population Research and Census 1992).

In the New Jersey system each of the five types of central places is identified by the nature of growth to be expected, the services to be provided, and *the connections to other places*. It is, therefore, a system of communities in which growth may occur. Douglas County, Oregon, uses the term "communities of place," although the emphasis on a system of communities is missing and the nature of growth is not an element. The county has identified three categories for unincorporated places—urban centers, rural centers, and rural settlements that do not meet a center criterion. The urban centers have at least five out of seven services available (water supply service, sanitary sewer service, fire protection service, public school, post office, community center, or grange hall) and all have sewer and water service and fire protection. The classification criteria for rural centers are 1) that the place have at least two out of the set of seven potential services, and 2) that it contain "urban land" defined as residential densities that are greater than one unit per 2 acres; sewer service; and residential and commercial or industrial uses (Cubic 1991).

Systems that control by specifying performance. This approach leans toward the market end of the spectrum. The term "performance" has acquired so many meanings it is necessary to state how we use it here (Porter et al. 1988). The distinction important to this discussion about managing markets and public and private investments is the relation between government and the property owner or developer. Standard land use control systems specify what a property owner must do in the interest of addressing the side effects of market transactions (this is the idea behind spatial bounding systems above). Pure performance systems enable the property owner to make use-by-location decisions provided side effects are kept within specified acceptable limits.[6] Put another way, standard control systems specify uses leaving side effects implicit; pure performance systems specify maximum permissible side effects leaving use implicit. Performance systems allow the market to function more freely. Development decisions are more efficient in economist's terms through the incorporation of the side effects. The project comes closer to its true costs; others are not absorbing uncaptured costs of the development.

Sanibel Island, Florida, and the Pine Barrens in New Jersey are examples of largely performance-oriented systems. They focus on the ecology of the site as the beginning point rather than economic rights in property. Breckenridge and Fort Collins, Colorado, have performance systems based on a more traditional economic view of property. Medford, New Jersey, is a system that incorporates both and is based on Ian McHarg's idea of using land for its "intrinsic suitability" (McHarg 1969).

Houston, Texas, the last large city in the U.S. without zoning (although it has had many of the components of such a system through nuisance law and covenants), is leapfrogging the standard zoning approach and is in the process of adopting a hybrid system limited to five zoning categories. It is considering allowing other uses in those categories subject to compatibility (performance criteria, degree of care) with the zone's protected use. That is, single-family residential use would be protected from other uses by forcing compatibility with the single-family housing. This concept is similar to some of the accommodations in Oregon wherein development is allowed in resource areas subject to its compatibility with the continued resource use. Any use in a zone, except the one by which the zone is identified, is treated as a conditional use that may not impair the prime or protected use.

While by no means a prevalent form of land use control, performance zoning is being used within Oregon. Bay City (1990 population 1,005) is a rural community on Tillamook Bay on the northern coast. In 1978, the city adopted the first performance zoning ordinance in Oregon (Pease and Morgan 1980). The city of Ashland incorporates a performance approach in its subdivision regulations by specifying overall density and lot coverage with no minimum lot sizes or housing types. Density bonuses are given for a variety of publicly valued attributes including energy and water conservation and affordable housing. Street widths, sidewalks, and off-street parking requirements correspond to the subdivision's scale.

IV. Application to Oregon

Over the past twenty years, there have been changes in urban patterns across the United States, changes in patterns of land uses within urban areas, and a significant shift in community attitudes about the environment. Oregon is in the vanguard of environmental concern with a population that values the natural environment for its contribution to the quality of life, economically and ecologically. Oregon's easy access to natural environments has reinforced the nationwide trend for less urban-centered living. This can be observed in exception areas throughout the state. The Oregon state system of development management, however, may no longer be appropriately tuned to these changes. Suggestions for alternate perspectives follow, based on the understandings and frameworks presented in parts I through III.

The objectives that the designers of the state growth management system appear to have set for themselves are three-fold. First, they wish to preserve two unequal land markets—urban and production lands—side by side. In order to do so, they must prevent the physical invasion of one use into the other; hold down land values on resource lands, to eliminate speculation, or to ameliorate the effects; and prevent or ameliorate the negative functional side effects of the urban use on the natural resource use. The solution of UGBs treats the two areas spatially in static terms. Second, they wish the solution to be adaptable to change over the long term. The task here is to be able to move the fixed boundary (UGB) without engendering speculation. This is a dynamic view. Third, they wish to encourage local governments to provide for orderly development that is compact and grows outward from an edge. Put another way, they wish to have efficient provision of services and prevent or ameliorate the functional and land value side effects of any development not directly related to production lands.

The items most at issue in system design seem to us to be: 1) the treatment of land use categories in the system, particularly the scope of urban and natural resources; 2) the grand scheme or system relating the two categories; and 3) the specific techniques employed. We will address these three points, moving back and forth among them.

In part II we noted that government draws on two streams of authority and techniques in managing change—legal, especially regulation, and fiscal. The Oregon system emphasizes regulation. We will spend more time on fiscal techniques. This does not imply they are more important, but they need attention to balance the system.

The treatment of land use categories

Urban. The rural development issue arises in significant part because the urban category of the state system is narrowly limited to incorporated areas. On the other hand, the county proposals reviewed in part III take an opposite view. They tend to treat rural as a further continuum of urban, not a discretely different thing. We think a revision of the urban category in Oregon's statewide system should be considered. A more appropriate definition of urban than as "incorporated place" would be urban as in the geographer's notion of "human settlements," a concept which is used throughout the developing world. The revision should reflect both the changing patterns of human settlement and the functions these settlements perform in the economy and residential structure. Such a revision would move more of the exception areas, or at least their urban or urbanizing parts, into this more structured "human settlements" category.

To identify, designate, and plan for human settlements, the following questions would need to be addressed. What has been the pattern of change in settlements—growth or decline, shifts in function, changes in age or other significant social structure variable? What functions do they perform? How are the settlements related to each other? What support systems do they need to perform these functions and maintain these relationships? What is or should be their relationship to the natural resources?

Hart and Mayer (1991) note that the effects of growth in the Twin Cities could be seen in community population increases for a distance of 75 miles outward. This was the result both of direct commuting into the urbanized area (which extends well beyond the 75-mile limit) and also of the decentralization of manufacturing and other functions into the communities within this 75-mile orbit.

The vision of Portland given by the Oregon state development management system is of a single urban mass, to be corralled by its UGB. But is it engendering growth in outlying linked communities in the manner of the Twin Cities? If so, how are they or should they be related or connected? Further, what has been happening to all the other communities or networks of communities in the state? While the UGB may be the best way to control the spread of individual places, a broader human settlements framework is necessary to address the larger policy issues of linkage and clustering among communities.[7]

The New Jersey system is a useful starting point, though its analysis of function is somewhat weak. In Oregon, some urban places undoubtedly perform a support function to farming and forestry. Some are resorts tied to the natural resources in a different way. They may be retirement communities. Others are part of a larger, nonland-based economy in which the location is accidental or incidental to the function. Many would perform more than one or all of these functions.

But the existing system extends further. It also includes the isolated industrial plant, the low-density subdivision, and the lone homestead within a forest. In fact, the human settlements pattern extends across the urban, exception, and exclusive resource categories. How many of these less urban types of settlement are desired by individual residents; how many are desired by residents collectively as a matter of state policy? If desired by both, they should be recognized and facilitated. If not desired by the community because of adverse impacts on other uses, the settlements should be discouraged by an array of techniques discussed below.

Natural resources. As noted, the early view of natural resources in the United States had an economic orientation. Resources were there to be consumed for the benefit of settlers—sites for houses, fields for farming, forests for timber and wood products. Twenty years ago an emerging concern about degradation of the natural environment began to be reflected in national legislation. An environmental ethic had already developed in Oregon as indicated by the bottle bill law, protection of public access to ocean beaches, and the land use system itself. More recently, environmental concern has focused on Mother Nature's use of the land, on habitat for plant and animal communities, and on endangered species as indicators of ecological system health. Federal agencies, among others, are beginning to perform analyses in terms of ecological systems looking beyond the single site under consideration in an impact review and examining it as an integral part of a larger ecosystem.

The Oregon state system speaks to all these concerns in its various goals. But it does not contain a conceptual framework for addressing as land policy these most recent perspectives on natural resources (nor does any other state at this time). We briefly presented one such conceptual framework in part II. We do not argue for that as a model. Rather, as with the New Jersey approach to settlements, we offer it as a point of beginning.

The Oregon system emphasizes natural resources as productive economic uses. It also speaks to the control of environmental pollutants and environmentally sensitive areas. The need, now, is to create the next new advance in integrating traditional land use management (the economic use of land), management relating to ecological communities or habitats, and pollution management (the adverse effects of the first on the second).

We have identified performance-based systems that begin with ecological considerations rather than economic ones. The spotted owl issue is about the health of the ecological community; it has a spatial dimension. It is this linking of economic and ecological considerations spatially that is the next great challenge in land use management. Oregon is uniquely qualified to be the innovator in this area because of its environmental orientation, the importance of natural resources in its economy, and its leadership in land use management.

Like the urban category, Oregon's resource category has the "one-size-fits-all" problem. Some resource lands lack the productivity to make them economically viable. On the other hand, natural resource-based uses such as wildlife management could be successful, but are not specifically provided for or allowed in present exclusive farm or exclusive forest categories. The richness of the Willamette Valley does not extend across the whole state. Variant uses, other than by exception, may be appropriate in these less intensively settled areas that also have less productive soils with less income-generating potential.

The grand scheme or system strategy

As we have attempted to explain, urban and rural as a dichotomy no longer capture the meaning of the present land use pattern. Major firms now can operate from former rural service centers via electronic connection and vegetables are raised in greenhouses in urban centers. The old dichotomy, regardless of its nostalgic roots, contributes little to understanding and problem solving. In short, "rural" as a category is at least twenty years out of date.

But there is power in the simplicity and logic of a two-part system. It is appropriate for several reasons. One stated intent is to maintain a visual distinction between urban and countryside. Another, as in traditional zoning, is to have exclusive use categories in order to protect each use from other incompatible uses. The scheme does not capture the lifestyle question represented by exurban uses of exception lands. However, responding to this lack with a third category of "rural" could defeat the market-management strength of the current system.

No other state-level jurisdiction attempts this feat. Indeed, it is not even clear that the operators of the Oregon system understand this as their system's greatest challenge. An alternate system that controls development through an adequate-facilities ordinance could control the urban extension in its physical manifestation. But it would not and could not contain the urban land value extension in larger urban regions, although it likely would reduce it somewhat. A performance system would have even less effect on land values, as its prime purpose is not to control location except through unacceptable side effects. As agricultural land often is prime development land, it is not likely that a performance system would have any of the intended effects. However, a combination of capital extension and performance, which will be discussed below, could be more effective than either one alone, particularly in areas of low market pressure.

We believe a two-part urban/natural resource system with the definition of "urban" broadened to "human settlements," and "natural resource" reexamined to include other less economically productive uses of the resources could be a viable emendation of the current system. It would preserve the strength of the current two-part system in managing the market, while responding to the weakness of narrow definitions. It would avoid creating "rural" as a third category. And it could lead to the elimination of the exception zones through reclassification into one of the two main categories.

Emphasis on the two-part view, to this point, has been to respond to economics and markets, to see clearly in this regard what is and must occur for policy to succeed. But there are other reasons, equally as compelling. They relate to the environmental considerations facing local governments and to the increasing array of fiscal devices employed to manage development.

Henry George put it best in *Progress and Poverty* over a century ago (1879). George made the important distinction between "land," by which he meant all the natural resources, as a gift of Nature, and "development" used broadly to mean the result of human labor. In this framework, what we describe as economic activity—human settlements, farming, forestry—all involve human labor in addition to the gift of Nature in the basic resource. He went further to describe why the distinction was important. It was that the differences in land value were not the result of the efforts of the owner. Rather, three contributors create value in property. The first is the gift of Nature in site qualities such as soil fertility, annual rainfall, natural landscape, subsurface minerals, or virgin forests. The second is value created by others who invested and developed in the vicinity of the site in question. The growth of a community and its resultant demand for sites is not a product created by a site owner. Finally, investments by government on behalf of the community also add value to a site. George then proceeded to argue that, since none of this value was created by the owner, it should be seen as belonging to the community. In specific, he argued for private ownership of land with a land value tax that would claim the economic rent from land for the community. He further advocated no tax on improvements as they were the legitimate fruits of human endeavor and belonged to those who created such value and to tax them would discourage investment.

We do not wish to explore or evaluate this whole line of argument. However, this philosophical framwork is a solid grounding for discussing public intervention necessitated by the market failures discussed in part III. That is, *who created the values, who reaps the benefits, and who pays the costs?* In part, this kind of analysis has led to a recent spate of new exactions, fees, charges, and the like. The conceptual framework is different than that of standard economics which construes a use value and a development value for a site, seeing all sites as part of one market (Peter Wilson and Associates 1990). The point is important because Oregon is attempting to separate the two competing markets of urban development and natural resources. However, the system contains mixed messages. By designating the uses to exist side by side, it is stating that public policy dictates use, not the market. But by introducing preferential taxation or deferral of taxation of resource lands the system recognizes the speculative land value for development that overlays the actual resource use value. It treats this speculative value as legitimately belonging to the property owner, and compensates for the fact that the owner cannot really afford this value for a resource-based enterprise. In short, the state acknowledges that higher land values generate tax loads incompatible with the earnings of resource-based activity. In addressing this problem with tax forgiveness, economic theory would suggest the effect is to increase the value further. Thus, to solve the short-term cash flow problem of the resource land user, the tax system actually increases the value of the land. The larger framework provided by Henry George offers an alternate way to think about this issue. His framework also applies to environmental issues.

Combining the natural resources framework set out in part II with the needs of development might produce something like the following. Recall that we identified four functions performed by the natural resource base—economic surface for development, land as a resource used in production lands, natural processes such as aquifer recharge beneficial to humans, and ecosystems or the interconnected natural community of plants and animal life on which we depend (see page 250). These functions are not likely to be separate physical spaces, but aspects of every site. Some sites may encompass all four functions; some may include only one. Consider criteria for human settlements that combine sensitivity to these environmental functions with settlement support systems. For example, downtown Portland might be at one end of a continuum, requiring a limited amount of environmental sensitivity and a large commitment to support services. At the other end of the continuum, an estuarine zone would require a high degree of care for any use other than those of Mother Nature herself and a low- or no-commitment policy on development and support services. It seems possible to construct a scheme for human settlements that reflects the locally specific relationship between development and the way in which resources are used based on their sensitivities and economic values.

Systems and techniques

The prime techniques employed in the state-level system are the UGB and exclusive production lands categories, both regulatory devices. These are supplemented by farm or forest use preferential taxation and right-to-farm laws. As we noted, UGBs appear to have been reasonably successful to date, but they "leak." High-density uses are kept inside, but low-density uses escape the boundary (Knaap and Nelson 1992). Further, selected case studies indicate much development has occurred outside the boundary in some communities (ECO Northwest et al. 1991). As that external scatter increases or as an urban reserve area is identified, one would expect land values to rise, an indicator that expectations about conversion of other sites is a real possibility.

If the state combined other fiscal techniques with its current regulation, the strategy would be strengthened. Obviously, specific choices depend on social acceptance and may require new legislation, particularly as these devices would bump caps or limits on property taxes imposed by Measure 5. But we list some possibilities for consideration. While all these techniques have been used in various settings, we are not aware of their use precisely as suggested here. No other place has attempted to do what Oregon is doing as a statewide system—achieve the land value difference without buying property rights with money (purchase of development rights) or allowing them to be exercised elsewhere (transfer of development rights).

As space prevents discussing all land use relationships, we will focus on the major change recommended, expansion of the concept of the urban category to include all human settlements. This would span the spectrum from Portland to the single cabin in the forest. The definition derives not from toying with the idea of urban-ness, but would include all development that is not a farmstead or similar resource-related dwelling. We will limit ourselves to the criteria related to controlling functional side effects, land value side effects, orderly provision of support services, and adjustments over time.

1. At the end of the continuum where the Portland metropolitan area is situated, the four components in the state system are as described above—UGB, exclusive resource zones, preferential taxation, and right-to-farm provisions. As several of LCDC's Urban Growth Management Study reports have suggested, a more orderly extension of the UGB could be achieved through the use of an adequate-facilities ordinance tied to a capital improvements program that would delineate a serviced-area edge. That edge could then be used to terminate the preferential taxation for forest and farm land that as a matter of policy should now be urbanized. This linkage to the other elements would relieve some of the pressure on the UGB. But it would not address the question of paying for the infrastructure.

Preferential taxation enables the farmer to live with the cash flow levels of resource-based production activities. In certain instances, the sale of the property will require repayment of the tax benefit that accrued to the landowner. There is no recovery by the public sector of the value created by the development of others; this is the driving force behind speculation. As Peter Wilson and Associates (1990) demonstrated, the property tax at true market value would drive the land onto the market earlier. The preferential tax allows it to remain in resource-based use. As they also noted, this can mean sheltering beyond the point where the land should be developed. If the state wants these lands to remain in resource-based production use long into the future, the combined taxes that confer a benefit on the farmer or forester by reducing cash flow requirements arguably should be combined with a tax to discourage speculation or conversion. A stronger reinforcement of such a policy would follow the view of Henry George and recover the value created by others, or some portion of it. This land value tax could be accrued until sale for urban use in order that it would not force premature transfer. These revenues could be used to help fund the infrastructure extensions. In short, the regulatory system establishes the community's preferred uses of land. Tax and fee systems could reinforce these preferences more fully by addressing the windfall received by the property owner when land is converted or the pressure of taxes related to speculative land values. The key questions, as noted earlier, are *who created the values, who reaps the benefits, and who pays the costs?* Every attempt consistent with policy objectives should be made to see that values and costs accrue to those who create them.

If the idea of an urban reserve is to be used, consider coupling it with value capture. In this instance, all of the value rise when the line is moved is owing to the actions of government combined with the investments of other property owners. This argues for capturing the full increase between the value of production lands and that of urban land.

For a less complete capture of the value increase, consider an approach like the Vermont transfer tax. In essence, this taxes the gain which Henry George would argue belongs to the community. But rather than taking the annual "rent," it is a one-time charge based on the rate of rise of value. It increases with the amount of value increase and decreases with time of holding. A large gain over a short period would result in a high charge; a small gain over a long time would have no charge. Oregon may wish to consider a higher charge rate. Vermont, after all, was not attempting to halt development, just the rate of rise in raw land sales.

2. If the extended concept of human settlements is adopted, UGBs may be appropriate in large, high-growth-rate areas, but not be necessary in areas of slower growth or smaller scale. When growth is slow in rate or small in magnitude, it will have small speculative impact. Therefore, the need to create an impregnable barrier between urban and resource lands is less compelling. A combination of an adequate-facilities ordinance with performance criteria may be a workable alternative. Performance approaches need sharply stated measures and they need oversight. But it is possible to tune to many more factors with them than with standard zoning classes. What cannot be accomplished by performance measures, however, is a blocking of land value increases. As noted earlier, some form

of value capture related to the capital extensions could depress speculation and also generate revenue for the infrastructure.

Settlements of larger size or rate of growth, between this mid-scale category and the Portland metropolitan area, would at some point shift to using the UGB. With the research behind the adequate facilities ordinance and capital programming, the perimeter should be reasonably easy to establish. Settlements of smaller scale or rate of growth could have even simpler criteria closer to the pure performance end of the continuum described next.

3. At the isolated development end of the scale, a tuning of the performance system currently used for nonresource dwellings in forest lands may be appropriate. One of the effects of scattered development in exception areas and in forests, for example, may be to enhance land values. As Knaap and Nelson (1992) made clear, land values tip up on both sides of the UGB, perhaps because of the access to open space created for those inside the UGB and the access to services for which no tax was due for those in exception zones. Are there ways in which these benefits consumed but not paid for can be charged? True costing of services would be a start, especially for those in exception zones who probably have the greatest free ride.

With these examples, we have attempted to sketch out possible additions and emendations to the existing system to sharpen its capacity to manage the new urban form. We have not detailed the possible techniques for the new environmental circumstances in which all of the United States finds itself, nor do we feel qualified to do so. As we noted earlier, that is the prime challenge that lies ahead.

V. Closing Thoughts

We have said numerous times that the Oregon system is unique in its attempt to enable two competing (and unequal) land markets to exist side by side—urban lands and production lands. That means that many of the strategies and techniques will be untried in these circumstances. Oregon will have to continue to pioneer.

The notion of human settlements puts rural communities on par with cities. In so doing, the "land between" once again disappears. The two-category urban/natural resource system is retained in essence, although it would be more appropriate to rephrase the dichotomy as "settlement/ natural resource." Giving credibility to rural settlements also makes it possible that over time these areas could emerge as centers of growth. This could take some pressure off existing metropolitan areas to continue growing outward in an undifferentiated manner. But the commitment to a two-part, two-market schema for growth management means that the remaining elements in the human settlements spectrum must be as carefully related to natural resources as the UGB. The means may vary, but the ends must remain the same.

The desire to live in rural settings is not going to go away. Polls taken for many years have always shown a strong preference for living in rural settings (Beale 1988). The 1970s saw a major movement to rural areas. Although this trend abated in the 1980s, it is more complicated than the aggregate numbers indicate. Retirees continued to move to nonmetropolitan areas (Butler 1991, Beale 1990). Research shows that this movement is conditioned on availability of jobs (Beale 1988); the greatest demographic change correlated with change in the sector of the economy in each county. As job conditions improve with further upturn in national and regional economies, the preference for moving to rural areas again may become a pressing reality. Oregon should be prepared.

Notes

1. When agriculture is intensive as in Florida's citrus groves and Holland's flower-bulb fields, it can often generate greater returns than low-density housing.
2. The tax deferral lessens the land value effect on cash flow, but not the impact on resource-based investment. For that, confidence of long-term continuity is needed. There is some question whether the recently approved urban reserve rules will reduce this confidence in the fixity of the urban land value line and cause resource land values to rise.
3. Space does not permit opening the further aspect of this issue—the appropriate method of accounting for cost-benefit analyses. As Repetto et al. (1989) have noted, present analyses are analogous to cash flow calculations. They include income (or its loss) and expenses incurred. They do not include change in asset position before and after the action analyzed. Therefore, much damage to natural resources is a consequence of converting assets to income, not truly increasing wealth or welfare.
4. The purpose of these categories is to state the functions nature performs for us, and the criteria for decision making. If multiple categories exist on one site, development uses have to meet multiple criteria.
5. A more comprehensive treatment of this relationship would add the role of nonprofit or third-party organizations. Generally this role is small in the type of growth management described here. The role of 1000 Friends of Oregon is significant, but basically it is to reinforce or clarify relationships and actions in the law and its implementation. In some settings, third-party organizations create action through initiative and referendum and constitute a distinctly different force.
6. Confusion occurs today when the term "performance zoning" is used to mean design specifications for mitigating elements (Kendig 1980). Standard zoning ordinances often include design standards that specify side yards, fencing, and landscape features of specified width. In this case, the property owner has no options to meet the requirement; it is specified. In contrast, a performance ordinance would specify the maximum traffic, glare, noise, and the like that could occur at the adjacent residential lot line, giving the developer latitude in meeting the specifications. Performance zoning as used here does not encompass the more rigid design standards used in standard zoning.
7. This work need not require a great research effort, however. A structured one-day workshop with knowledgeable professionals probably could create a good approximation of how the state settlement system functions.

References

Alonso, William. 1964. *Location and Land Use: Toward a General Theory of Land Rent.* Cambridge: Harvard University Press.

Anding, Thomas L., et al. 1990. *Trade Centers of the Upper Midwest: Changes from 1960 to 1989.* Minneapolis: Center for Urban and Regional Affairs, University of Minnesota.

Beale, Calvin L. 1988. "Americans Heading for the Cities, Once Again." *Rural Development Perspectives* 4, 3: 2-6.

———. 1990. "Preliminary 1990 Census Counts Confirm Drop in Nonmetro Population Growth." *Rural Conditions and Trends* (Winter).

Butler, Margaret. 1991. "Rural Population Growth Slows during 1980-90." *Rural Conditions and Trends* (Spring).

Center for the New West. 1992. *Overview of Change in America's New Economy.* Denver: Center for the New West.

Center for Population Research and Census. 1992. *Population Estimates for Oregon 1980-1991.* Portland: Portland State University.

Christaller, Walter. 1933. *Die zentralen orte in Suddeutschland.* Jena: A summary in English is found in Rutledge Vining, "A description of certain spatial aspects of an economic system," *Economic Development and Cultural Change* 3, 2: 16-147.

Cubic, Keith L. 1991. "Rural Communities Petition" to Land Conservation and Development Commission.

Davis, Judy S. 1990. Exurban Commuting Patterns: A Case Study of the Portland Oregon Region. Ph.D. Dissertation, Portland State University.

ECO Northwest, David J. Newton Associates, and MLP Associates. 1991. "Case studies report." A report in the Urban Growth Management Study. Salem: Department of Land Conservation and Development.

Einsweiler, Robert C., et al. 1975. "Comparative Descriptions of Selected Municipal Growth Guidance Systems: A Preliminary Report." In Vol II of *Management and Control of Growth: Issues, Techniques, Problems, Trends.* Edited by Randall W. Scott, , David J. Brower, and Dallas D. Miner. Washington: Urban Land Institute.

Garreau, Joel. 1991. *Edge City: Life on the New Frontier.* New York: Doubleday.

Giuliano, Genevieve, and K. Small. 1991. "Subcenters in the Los Angeles Region." *Regional Science and Urban Economics* 21 (July): 163-182.

George, Henry. 1879. *Progress and Poverty.* Reprint 1990. New York: Robert Schalkenbach Foundation.

Gleeson, Michael E., et al. 1976. *Urban Growth Management Systems: An Evaluation of Policy-related Research.* Chicago: American Society of Planning Officials.

Godschalk, David R., David J. Brower, Larry D. McBennett, Barbara A. Vestal, and Daniel C. Herr. 1979. *Constitutional Issues of Growth Management.* Revised ed. Chicago: Planners Press.

Hall, Peter. 1991. "Cities and Regions in a Global Economy." Working Paper No. 550. Berkeley: Institute of Urban and Regional Development, University of California.

Hart, John F. 1991. "Tough Times for Minnesota Small Towns." *Center for Urban and Regional Affairs Reporter* 21, 4.

———, and Tanya Bendicksen Mayer. 1989. "Small Towns Can't Stop Growing." *Center for Urban and Regional Affairs Reporter* 19, 2.

Kendig, Lane. 1980. *Performance Zoning.* Chicago: Planners Press.

Knaap, Gerrit, and Arthur Nelson. 1992. *The Regulated Landscape: Lessons on State Land Use Planning from Oregon.* Cambridge: Lincoln Institute of Land Policy.

Ladd, Helen, and William Wheaton. 1991. "Causes and Consequences of the Changing Urban Form." *Regional Science and Urban Economics* 21 (July):157-162.

Little, Charles E., and W. Wendell Fletcher. 1981. "Buckshot Urbanization: The Land Impacts of Rural Population Growth." *American Land Forum* 2, 4.

Logan, Thomas H. 1976. "The Americanization of German Zoning." *Journal of the American Institute of Planners* (Oct 76).

McHarg, Ian L. 1969. *Design with Nature.* Garden City, NY: Natural History Press.

Muth, Richard F. 1969. *Cities and Housing: The Spatial Pattern of Urban Residential Land Use.* Chicago: University of Chicago Press.

New Jersey State Planning Commission. 1988. *Communities of Place.*

Pease, James R., and Michael Morgan. 1980. "Performance Zoning Comes to Oregon." *Planning.* 46, 8: 22-24.

Porter, Douglas R., Patrick L. Phillips, and Terry J. Lassar. 1988. *Flexible Zoning: How it Works.* Washington: Urban Land Institute.

Repetto, Robert C., et al. 1989. *Wasting Assets: Natural Resources in the National Income Accounts.* Washington: World Resources Institute.

Schwartz, Joe. 1990. "Catch-22." *American Demographics* 12 2: 10.

Smithsonian Exposition Books. 1979. *The American Land.* New York: Norton.

Wilson, Peter, and Associates. 1990. "Property tax deferral policy inside urban growth boundaries." A report in the Urban Growth Management Study. Salem: Department of Land Conservation and Development.

CHAPTER 13
A Research Agenda for Oregon Planning: Problems and Practice for the 1990s

Deborah Howe

Research Experiences

The Oregon land use planning system is widely recognized for its comprehensiveness, durability, and innovation. It is ironic, however, how little is known about the program's impact, effectiveness, and implementation experiences. The design of the program did not include a parallel and complementary research component and scholars have filled this need to only a limited extent.

One of the greatest shortcomings of the program is the absence of a database on land use patterns and changes. Geographic information system technology was not widely available when the program began in the early 1970s and thus it was not feasible to make long-term commitments to computerizing maps and related land use data. Of perhaps greater relevance in explaining the lack of a database was the intense political pressure to get local plans through the acknowledgment phase. What was envisioned to be a one-year task took ten years as policy makers and planners struggled with applying goals and policies at the same time that they were being defined and refined. There was little interest in the design and implementation of information systems to facilitate monitoring and evaluation.

In 1980, however, the Department of Land Conservation and Development (DLCD) hired consultants to prepare an evaluation design

(Economic Consultants Oregon, Ltd. 1980). Funded through the Coastal Zone Management Act of 1972, the proposed framework actually addressed most of the state planning program. It built on the notion of defining objectives and measurable indicators for each of the statewide goals and described various approaches to collecting the necessary information. Unfortunately, Oregon was experiencing a recession and the state had serious budgetary limitations when this report was completed. The Land Conservation and Development Commission (LCDC) was reluctant to undertake the rigorous value clarification and public declaration of policy directives which were needed in order to conduct an evaluation (Niemi 1991).

In response to the pending completion of all city and county plans, DLCD contracted with another consultant in 1983 to design a system for monitoring comprehensive plans (Richard L. Ragatz Associates, Inc. 1983). The resulting product was intended to provide assistance to local governments in setting up a procedure for collecting planning and land use data for use in improving the implementation process. Although a manual was ultimately distributed to all counties, no additional resources were provided to assist in implementation and the existence of the report was largely forgotten.

Four years later, the Bureau of Governmental Research and Service (BGRS) at the University of Oregon received a grant from DLCD to develop an evaluation design for Oregon's planning program (BGRS 1987). The intent of this grant was for BGRS to not only develop a proposal, but to actively seek funding to conduct the evaluation. The bureau assembled a team of Oregon academics to define a series of research strategies for addressing different facets of the program and brought national land use experts to Oregon to critique the proposal.

The initial cost of the evaluation was estimated to be $1.9 million; this was ultimately reduced to $1.6 million. The bureau sought funding for the project in 1987 and 1988 but was not successful. When the 1989 state legislature allocated $500,000 for two studies—on farm and forest lands and urban growth management—the BGRS proposal was not pursued further. The impetus for these two studies was a recognition that a number of mid-course corrections had to be made in the Oregon planning system. It was felt that the research would help in defining policy alternatives.

Participants in and observers of the farm and forest lands and urban growth management studies have had mixed feelings about the manner in which the research was managed and the actual results. The effort was welcomed; professionals, policy makers, and academics were hungry for solid information about the program and the studies filled a vacuum. However, a limited budget, severe time constraints, and political realities meant that these studies could only scratch the surface of what was desirable to know.

The research was done by contractors under the coordination of two DLCD staff members hired specifically for this purpose. Because of delays in administering the request for proposals process and changing expectations about final products, contractors had six to eight months to do the bulk of their work. The lack of readily available data necessary for the completion of the required tasks was a major constraint. Local governments did not have a good system for collecting data with most data being collected at a single point in time for a specific purpose. This type of database has limited value for research purposes.

Some of the consultants were reluctant to draw policy implications on the basis of their research results (Hope 1991). There was also concern that there was too much political influence in the research process, since interest groups were involved in framing questions and defining research methods. Several researchers felt that more discretion should have been given to the contractors and that DLCD should have insulated the contractors from interest group pressures. DLCD did not feel free to do this, however, since the legislature had required the agency to integrally involve interest groups (Howe 1991b). The interest groups welcomed the role they played. Planning professionals, in particular, appreciated the openness of the process. This helped to smooth over those situations in which disagreements could not be resolved (Childs 1991).

The summary report to the 1991 legislature included recommended policy responses. Implementation alternatives were to undergo additional review and analysis by various task forces in order to encourage a broader discussion and more in-depth examination of specific issues. The followup work program, however, was severely limited because the legislature did not provide the necessary resources. The department has looked at alternatives for developing a research program (Howe 1991b), but in the absence of funds and with changes in agency administration, there is no clear sense of what will happen.

Other sponsors of research about the Oregon planning program have included interest groups and academics. 1000 Friends of Oregon is in the forefront of interest group research. They have conducted a variety of studies on such issues as housing affordability, open space, land use planning and economic development, hobby farms, Land Use Board of Appeals activities, and dwelling permits in farm and forest zones. These studies are intended to further 1000 Friends' positions; their senior planner honestly admitted that the results would not be published if they did not support the point that the organization was trying to make (Ketcham 1991).

Academic researchers have privately expressed reservations about the tendency of 1000 Friends to use a short time frame for their studies, hence limiting the data that can be collected. There has also been concern with the lack of relationship between reported facts and conclusions. There is no forum, however, in which the organization's research methods can be critiqued. Because they are a private interest group, their work is not likely to receive much scrutiny because they, along with everyone else, are entitled to present their position as they see it. The problem in this case is that 1000 Friends is the primary generator of information about the Oregon planning program and therefore the information they provide tends to become the common wisdom.

Research by other constituent groups has been limited. Since the mid-1970s, the Portland Metropolitan Homebuilders Association has collected information on permit and subdivision activity, which they have shared with 1000 Friends in a housing affordability study. 1000 Friends has tended to take the lead in this type of collaborative effort because they have the research capacity (Hales 1991).

Academic research has been dependent largely on shoe-string budgets. This has resulted in the use of data sets that are limited in size and geographic diversity. Lack of resources has caused most attention to be directed at program impacts in the Portland metropolitan area and the highly productive Willamette Valley. Conclusions drawn on the basis of these studies can mask the differential impact of the program in other parts of the state.

Most of this research has been conducted by scholars from outside the state. Oregon academics have not collectively developed a strong tradition of research focusing on the state planning program. As a result, the research that exists does not benefit from the intimate knowledge of the political, economic, and social context that Oregon academics could provide. Of perhaps greater significance, Oregon academics have failed to coalesce as an interest group that can effectively support DLCD's efforts to develop a research capacity.

What research exists has not been effectively accessed by decision makers. Planners and academics are routinely frustrated by the extent to which key policy choices are made on the basis of anecdotal evidence rather than research findings. The player who has the best story wins. This situation may be explained in part by the schism that occurs between practitioners and academics. Academic publications are generally inaccessible to practitioners and research results often have little direct relevance to the person working in the trenches. On the other hand, practitioners frequently neglect to look beyond their direct experiences to seek a broader context and new insights. Thus it should not come as a surprise that the work of the academic researchers is virtually unknown among Oregon planners.

Developing a Research Agenda

Research about the Oregon planning system has two potential consumers. Within the state, it can serve as a basis for making improvements in the implementation process. For those outside the state, research can provide insights to guide the development of other planning innovations. Because Oregon has not created a climate supportive of research, potential benefits for improving the program are not being realized. This in turn limits the usefulness of the research to outsiders because what is available is underfunded and reflects in only a weak sense what could be learned within an environment that is more constructively critical and self reflective.

Thus, in developing a research agenda, the crucial task becomes not the identification of research topics, but rather the fostering of a more supportive culture. This is a multi-faceted challenge that involves encouraging practitioners to have a research orientation, demonstrating the relevance of research to decision making, and making better use of existing research. It is not necessary to wait for the big dollars before research can become a priority. In fact, extensive funding is not likely to become available until the value of research is apparent.

There are a few promising steps in these directions. The Oregon Progress Board (1991) *Benchmark* report to the 1991 legislature is being used as the basis for measurable standards for evaluating and funding state agencies. This report is a compilation of quality-of-life indicators with specific dates by which improvements are to be achieved. Agencies are now challenged to focus their efforts to address these standards. In so doing, they are confronting the relevancy of the measure. DLCD, for example, realizes that the percentage of development within urban growth boundaries (UGB) says nothing about the quality of that development. The resulting discussions are helping to highlight the importance of research in public decision making. In addition, the DLCD studies, while a source of frustration to many, underscore the lack of readily available information about the program. The obstacles encountered in doing these studies and the implications of a lack of information can be used to build a case for more funding support to address the remaining research questions.

Institutional Alternatives

In a survey of practitioners and academics, there was widespread support for the idea that DLCD should define a stronger research role (Howe 1991a). There were different opinions, however, as to how this should be accomplished. Suggestions included housing the research function within a university, devoting one to two DLCD staff members exclusively to research, and creating a research/forecasting unit outside of DLCD to articulate a long-term perspective and to work on crosscutting issues relevant to various state agencies. DLCD has been encouraged to create an advisory board to help in defining a research agenda. There has also been a call for development of an institutional mechanism for peer review of research including study design, findings, and conclusions.

While the resources for realizing these ideas are not likely to be available in the near future, the need for credible, useful research will continue to grow. This underscores the importance of making the most of what is available. Top priority should be given to the development of a database as a part of local monitoring. Readily accessible, good-quality data will go a long way toward encouraging the simpler, less costly studies that can be effectively done by scholars, graduate students, and interest groups. A well-structured database can also be a key factor in seeking funds for much larger research initiatives.

A concerted effort should be made to define data needs and research questions. All studies prepared or sponsored by DLCD should include as a matter of course a discussion of further areas of inquiry including a consideration of their priority and what data would be needed. Interest groups, planners, and academics should be regularly asked for this type of input. A research advisory board could review this material and send out a call for research.

A research basis for all major policy issues should be regularly developed. This would not necessarily involve collecting new data. It could consist primarily of a review of existing research, something that could be done by a graduate student at minimal cost. Objectives should be clearly stated with measurable criteria. This will facilitate evaluations, underscore the relationship between research and policy making, and make policy alternatives clearer.

The dialog among practitioners, interest groups, policy makers and academics about the appropriate way to do research can be constructive if a collective effort is made to understand and appreciate differing perspectives. Consultants may have more experience with less complex policy studies. Academic researchers may be more comfortable with developing a more rigorous, defensible database before proceeding to policy formulation. The consumers of the final product want something that is directly useful. If DLCD develops a better understanding of these perspectives, it would be possible to take greater advantage of the different research capabilities. The researchers, in turn, would also benefit from learning how their work can be more useful in actual application. In any case, it is important that a dialogue enhance support for research rather than discourage future efforts.

The Oregon program, while innovative, does not have a mechanism for critically engaging new ideas. As a result, many people become frustrated with what seems to be overwhelming system inertia. A research perspective builds on the basic premise that there is something to be learned through thoughtful questioning, analysis, reflection, and interpretation. At that point, the door is opened for creative concepts that could allow the program to more fully realize its potential.

Defining a Research Agenda

The research that needs to be done falls under four interrelated categories: monitoring, evaluation, applied research, and visionary research. We will consider each separately.

Monitoring

Monitoring is the systematic collection and reporting of specific data in order to determine what is being done in implementing a program or policy. While the underlying intent of monitoring may be evaluative in nature, there are some important differences. Monitoring tends to have a fairly narrow focus on particular policy concerns. Unless the users of the data look critically at what is being learned about program implementation, the benefits of monitoring will be limited.

A database does not constitute a monitoring system without the reporting component. The audience for monitoring can include, among others, program administrators, policy makers, clients, or the public at large. In Oregon, the state legislature has required DLCD to report regularly on dwelling activity on resource lands. On its own initiative, the department reports on Land Use Board of Appeals activity. At the local level, a planning department might set up a system for tracking variance requests. This system could serve a monitoring function if the results are regularly reported. This example indicates how a monitoring system can be incorporated with an information system that is an integral part of program implementation.

As part of the periodic review process, counties and cities should be required to set up a monitoring system that yields information that is useful to the local government and can be aggregated for regional and statewide analyses. The type of data that should be collected in both mapped and tabular form includes subdivision activity, cross-referenced with plan map and zoning designations and proposed/existing public facilities; building permit locations, building type, and construction costs; conditional use and variance requests as well as zoning and plan map amendments; dwelling units permitted and denied in areas designated for resource use; and parcels that are receiving agricultural tax assessments. The information system should also track land use activities from application to disposition.

A monitoring system needs to respect the resource and time constraints of local governments and its usefulness and value should be readily apparent. It would be preferable to have a system that could be adapted to both computerized and noncomputerized environments. If the system is to be used to develop an understanding of statewide land use issues, then the data need to be consistently defined, collected, and reported. Involvement of researchers and practitioners in defining the components of a monitoring system will help ensure that it meets the needs of different users.

Monitoring serves an important function in enabling administrators and policy makers to keep an eye on a program. Severe deviations from what is expected to occur can flag situations in which mid-course corrections in administrative procedures or perhaps more in-depth evaluation and analysis are appropriate.

Evaluation

Evaluation involves an assessment of the extent to which a program is successful in accomplishing specified goals and objectives. The data generated in monitoring can be useful in program evaluation if enough forethought is given to the design of the system. Evaluations are often done before a program is implemented in order to develop a sense of whether the proposed program will significantly alter existing trends and whether the effort will be cost beneficial. New Jersey's impact analysis of their proposed state plan represents this type of pre-program evaluation (Rutgers University 1990). It is more common for evaluations to be done on a *post hoc* basis, after a program is well under way. When this type of evaluation is undertaken without any preplanning in advance of program implementation, significant challenges will likely have to be addressed because of the absence of useable data and, perhaps more importantly, lack of clarity about the original goals and objectives. This accurately describes the problems inherent in evaluating the Oregon program.

There is another complicating factor—the context has changed. Oregon's program started in response to severe development pressures that were threatening the state's farming and forest industries. Two decades later, the forest industry is in decline, with serious repercussions for timber-dependent communities. The state's economy is globalizing with the growing importance of Pacific Rim trade. The population is becoming more diverse through aging, in-migration of minorities, and other significant demographic changes such as the increased number of single-parent households. An evaluation could reveal that the program was an adequate response to the original needs, but, contextual changes might make the program irrelevant in meeting current and/or future needs. Michael Hibbard's research (chapter 9) has indeed indicated that the program by itself may be an inadequate tool for addressing the current development needs of many smaller, single-industry communities.

Not everyone agrees with the call for program evaluation. A planner for the Metropolitan Services District felt that "at some point we should declare victory and move on" (Seltzer 1991). There is some validity to this perspective in that the true impact of the program in containing growth and protecting resource lands may not be apparent for several generations. Nevertheless, there are issues that merit evaluation research in order to maintain the vitality and relevance of the statewide planning program.

Concerns raised by Robert C. Einsweiler and Deborah Howe (chapter 12), Michael Hibbard (chapter 9) and James Pease (chapter 8) underscore the importance of giving priority to rural lands and rural communities, a source of festering concern for the program. It is widely perceived that the Willamette Valley perspective dominates the program. The resulting policies that derive from the valley's high population concentration, high growth rate, fertile farm land, and abundant rainfall are seen as being problematic in the much larger "other Oregon" which has sparse population, poor soils, and low rainfall. In central Oregon, exclusive farm use (EFU) zoning is leading to the creation of specialty farms (such as llama husbandry and Christmas tree farming) as a means of legitimizing dwelling construction. One planning director has privately expressed concern that the EFU policy forces land into inappropriate agricultural uses which compete with wildlife, industry, and urban settlements for scarce water resources. He feels that the program should recognize open space preservation as a legitimate land use and provide for a dwelling as a permitted use. But he fears making this suggestion as it runs counter to the statewide planning goals.

An evaluation of rural lands policies needs to begin with a systematic determination of the prevailing issues and how these vary from one part of the state to another. In other words, this study should recognize and articulate the state's inherent diversity. There is a need to develop an understanding of the interplay between the planning program and a variety of variables including climate, water, and soil; federal rangeland, forest, river flow and endangered species policies; and the nature and structure of resource related industries. Einsweiler and Howe's concern with the relationship between the planning system and differentials in land markets could provide a framework for this analysis (chapter 12).

Consideration also needs to be given to institutional aspects of program implementation. Ed Sullivan sheds light in chapter 3 on the administrative detail required by the program. Although this is not strictly a rural issue, the ability of different jurisdictions to comply with state planning mandates, particularly in light of limited budgets is a concern. It would be helpful to know to what extent the increasing legal requirements of the planning program hinder effective planning. Some requirements may be viewed as little more than busywork in depressed communities such as those analyzed in chapter 9. In light of the way in which the program was used to hinder the development of Rajneeshpuram, it would be appropriate to ask to what extent the program squelches innovation.

The adoption of the transportation rule provides a rich opportunity to evaluate implementation of a specific policy. LCDC's stated intention to reexamine the vehicle miles travelled goal every five years is a step in the right direction. Sy Adler suggests in chapter 6 several other concerns that merit attention, including the relationship between the transportation goal and other goals such as affordable housing and preservation of open space; the extent to which the market responds to the transportation goal; and the impact of neighborhood opposition on goal achievement.

There also needs to be an evaluation of the relationship between the state planning program and social issues. While Nohad Toulan (chapter 5) sees the state's housing policies as a powerful statement on government's responsibility to ensure the provision of affordable housing, the extent to which this value is effectively implemented is not clear. It would be helpful to define urban and rural housing needs and to assess how the housing supply is influenced by the planning program. With regard to other social concerns, it should be noted that the statewide goals do not currently take a proactive stance on the provision of child and adult care, on encouraging diversity in housing type to meet the needs of an increasingly heterogeneous population, and on planning for an aging society. If and how these needs are being met should be evaluated.

Applied research

The program needs to be improved, refined, and strengthened without always knowing whether the basic program parameters are making any difference. Thus the type of studies that are needed the most are policy specific with a strong emphasis on defining appropriate alternatives. This is the nature of applied research. A variety of policy clusters representing the program's response to addressing specific issues need attention. These include urban, exurban, and rural development; resource conservation; and agricultural preservation.

Urban development—What are the policy options for redeveloping underdeveloped areas? How can density increases be accomplished without destroying existing neighborhoods? To what extent and how can market demands for low-density development be discouraged?

Exurban development—What is the relationship between development within a UGB and exurban development on exception lands just outside the boundary? To what extent does exurban development undermine the fiscal solvency of urban areas? How do the travel patterns associated with exurban development affect regional demands for infrastructure and services?

Rural development—How much does it cost to provide services such as schools, fire suppression, and medical assistance to households living in a dispersed development pattern? How are management costs for the forest and agricultural industry affected by parcelization and the presence of nonresource dwellings? To what extent can both current and future rural development needs be adequately met through designated exception areas and small incorporated cities? What are the appropriate performance standards for enabling differing land uses to be in close proximity?

Resource conservation—In light of various efforts to protect endangered species, limit grazing on public lands, and reduce pollution, what are the costs of conservation and who bears these costs? How can these resources be developed and used?

Agricultural preservation—What are the disamenity effects of urban development on farm operations? Does the impact vary by density, type of development, and distance? To what extent do hobby farms support or undermine a commercial agricultural industry? If they are compatible, what should be the minimum lot size of the hobby farm? What is the environmental impact of farming? What is the relationship between farming, wildlife management, recreation, and tourism? How does the scarcity of water affect this relationship?

Applied research also involves seeking out, adapting, and developing planning innovations. Other states have historically looked to Oregon for inspiration. As these states have developed their own programs in response to their unique circumstances, they have created new approaches that can and should be a source of fresh ideas for Oregon. Oregon is now in a position to learn from others. It is important that the state keep abreast of what is happening elsewhere in order to maintain the vitality of the program. DLCD undertook a scan of other programs, looking for ideas on such issues as regional review of local plans, state agency coordination, concurrency, and guiding exurban development (Howe 1991a). This type of search needs to be done on a regular basis.

There is also a strong need for field testing, assessing, and disseminating planning concepts that can be put into practice by local governments. Zoning models for guiding neotraditional developments, site planning standards for special-needs housing, alternatives for managing wetlands in urban areas, and techniques for effectively involving citizens in the planning process are a few examples of the type of information needed by local governments.

Visionary research

Ask an Oregon planner to list all nineteen goals and you will find few who can do so. The reason is that the number of goals exceeds the memory capacity of most people (Mandler 1967). Thus the goals in their current form fall short of serving as a vision for the state as a whole. The high number of goals may also explain why some goals have become high-threshold concerns, receiving a great deal of attention, and others, such as the Energy Conservation goal (Goal 13) and Historic Preservation goal (Goal 5), have been relegated to the back burner.

The goals could be reorganized without loss of content. A nested hierarchy based on three categories is most conducive to retention (Miller 1956). One approach would be to group the goals in terms of substantive coverage. This might include decision-making process, resource management, and human settlements. The current goals would then be sorted among these categories. Some goals may need to be divided. Goal 5, Open Spaces, Scenic and Historic areas, and Natural Resources could be divided between the resource management and human settlement categories. Specific attention needs to be given to the choice of category titles. "Decision making" suggests the usefulness of planning and hence may be more understandable to a broader cross section of people than "planning." The word "management" incorporates the notion of *both* conservation and development. "Human settlements" responds to the ideas raised by Einsweiler and Howe (chapter 12) to ensure that unincorporated communities have a stake at the planning table.

An alternative approach would be to regroup goals to emphasize their interactive relevance for different parts of the state. This might result in a somewhat different constellation of goals for the Portland metropolitan region than for the more rural parts of eastern Oregon. This concept, however, would probably entail substantive changes in goal contents because the program is currently based on the premise that the goals are for the most part uniformly applicable throughout the state.

Reorganizing the goals would involve a systematic and careful study of goal content and intent. It would be an appropriate time to assess what is missing, such as a focus on social concerns. Reordering could lead to a change in emphasis, giving open space and wildlife management, for example, as much attention as agriculture and forestry.

Consideration also needs to be given to developing a vision for the state of Oregon that defines where growth should be accommodated. Looking back sixty to seventy-five years, one would find cities that at the time seemed poised to continue as dominant centers. But economic and technology changes and in some cases massive destruction due to fire meant promises unfulfilled. The city of Astoria, which had a population of 24,000 in the early 1920s, today has fewer than 10,000.

And so it would be erroneous to assume that today's development patterns are the only determinant of what will or should exist several generations from now. Oregon's planning program is somewhat conservative in its emphasis on concentrating development in existing cities. While new towns and satellite centers are not prohibited, they are not aggressively pursued as a planning policy.

Research is needed on defining growth potential and exploring the implications of economic trends and changing technologies with respect to where and how new development can be accommodated. The revolution in communications technology makes the decentralization of many businesses feasible. This in turn could be the focus of economic development strategy for many of the state's rural communities. In identifying where growth should be encouraged, consideration should be given to what public policies would make it happen. This statewide perspective could then serve as the context within which counties and cities could do their more localized planning.

Developing a vision for the state is not an easy task. In 1987, the Oregon chapter of the American Planning Association developed a process and undertook an effort to develop a vision, but they encountered a considerable amount of inertia. People had a difficult time considering the state as a whole. The chapter subsequently decided to refocus their efforts to the community level. Their hope was that if enough communities developed visioning skills the next logical step would be development of a statewide vision.

This visioning initiative, however, is originating with a private organization. If the state were able to underwrite the analyses that dimension the growth challenges that lie ahead, then a statewide effort could gain momentum.

Cutting edge research involves a commitment to make the planning program the very best it can be. It involves a constant search for new ideas and a willingness to test them and to assess their effectiveness. Ultimately some approaches may not work and so it is important to have the capacity to backtrack as necessary. But it is the cutting edge research, with its emphasis on innovation, quality, and challenge, that will ensure that Oregon maintains its leadership role in statewide land use planning.

References

Bureau of Governmental Research and Service (BGRS), University of Oregon. 1987. *Evaluation Proposal for Oregon's Land Use Planning Program*. Salem: Oregon Department of Land Conservation and Development.

Childs, Jan. 1991. Interview by author, 2 August.

Economic Consultants Oregon, Ltd. 1980. *Evaluating the Performance of the Oregon Coastal Management Program*. Salem: Oregon Department of Land Conservation and Development.

Hales, Charles. 1991. Interview by author, 20 August.

Hope, Jim. 1991. Interview by author, 20 June.

Howe, D. A. 1991a. *Development of a Research Action Plan: Issues and Opportunities*, Portland: Center for Urban Studies, Portland State University.

———. 1991b. *Review of Growth Management Strategies Used in Other States*, Portland: Center for Urban Studies, Portland State University.

Ketcham, Paul. 1991. Interview by author, 24 May.

Mandler, George. 1967. "Organization and Memory." In *The Psychology of Learning and Motivation*, Vol. 1. Edited by Kenneth W. Spence and Janet Taylor Spence. New York: Academic Press.

Miller, George A. 1956. "The Magical Number Seven, Plus or Minus Two: Some Limits to Our Capacity for Processing Information." *The Psychological Review* 63, 2: 81-97.

Niemi, Ernie. 1991. Interview by author, 18 June.

Oregon Progress Board. 1991. *Oregon Benchmarks: Setting Measurable Standards for Progess*. Salem: Oregon Progress Board.

Ragatz, Richard L. Associates, Inc. 1983. *Comprehensive Plan Monitoring: Guidelines and Resources for Oregon Communities*. Salem: Oregon Department of Land Conservation and Development.

Rutgers University, Center for Urban Policy Research (CUPR). 1990. Response to Request for Proposal for the Impact Assessment Study of the New Jersey State Development and Redevelopment Plan. Piscataway, NJ: Rutgers University.

Seltzer, Ethan. 1991. Interview by author, 8 May.

Afterword

What lessons and conclusions can we draw from twenty years of state-guided land use planning in Oregon? What can other states learn from Oregon, and Oregonians from their own experience?

Perhaps most obvious is the positive reputation of the system. After twenty years, outsider observers still treat Oregon's program as a benchmark. As John DeGrove has described, Oregon's planning experiment was a source of ideas for a number of eastern states during the 1980s, with Maine and Rhode Island most closely following its model of state-local partnership. In addition, the Oregon system has helped to validate more specialized environmental planning efforts in which local governments have retained front-line responsibility to prepare and implement comprehensive plans but have been required to bring such plans into compliance with policies and guidelines set by state-level agencies. Examples include the New Jersey Pinelands Protection Act, Maryland's Chesapeake Bay Critical Areas Act, and the Cape Cod National Seashore.[1]

The system's ability to weather four statewide votes of confidence and a gubernatorial study commission during its first decade surely encouraged political leaders in other states to invest their own time and energy in crafting the "second generation" growth management plans of the 1980s. In the private sector, 1000 Friends of Oregon has provided a model for mobilizing citizens and community leaders into a permanent land use planning advocacy group.

The state of Washington is now recapitulating its neighbor's experience. Washington's new growth management legislation—a response to booming growth in the Puget Sound region during the 1980s—is very much like Oregon's Senate Bill 10 of 1969. Washington's local governments are required to develop comprehensive plans to take into account a set of statewide goals. However, the legislation has provided no administrative mechanism for assuring the quality and effectiveness of such local compliance. Early evaluations have shown wide differences in the ways in which local planners have defined and protected elements of the natural environment and met other planning goals. It is likely that Washington will find itself moving toward its own version of Senate Bill 100, with stronger procedures for systematically testing local planning efforts against state standards.

In substantial measure, those outsiders who have looked to Oregon as a success story are right. The Oregon system has been effective in doing what it was designed to do—control sprawl in the Willamette Valley by creating an institutional barrier between suburbanites and farmers. Urban growth boundaries (UGBs) have proven an effective tool for separating markets for agricultural land and urbanizable land. Although metropolitan Portland, Salem, and Eugene have grown outward, most of Oregon's agricultural core has been insulated from leapfrog subdivisions and peripatetic industrial parks. The Willamette Valley has yet to be paved over.

Another part of the success story has been the growing ability of the Oregon system to adapt to changing circumstances. Although consensus is still to be achieved, the Department of Land Conservation and Development (DLCD) has started to think about the treatment of unincorporated rural settlements. System flexibility is reflected in 1992 administrative rules that define "urban reserves" as a required tool for some but not all counties.[2] Similarly, the Land Conservation and Development Commission (LCDC) has fine-tuned Goal 10 by writing a housing rule specifically for the largest metropolitan area and exempting small cities and counties from the new transportation rule.

More importantly for other states, the Oregon planning program has produced some nationally significant innovations. Everyone involved in urban and regional planning, for example, recites the same litany about the inextricable connections between land use and transportation decisions. Coordinated planning, however, has usually been blocked by the cumulative differences in origins, funding sources, political allies, and institutional cultures of land use planning agencies and state transportation departments. In Oregon, the state planning system has helped to turn DLCD and the Department of Transportation into allies. The state's new transportation rule, as discussed by Sy Adler in this volume, tries to bridge the chasm by a travel reduction goal—a 20 percent decrease in vehicle miles traveled per capita—that local governments can hope to meet only by linking transportation and land development decisions.

Equitable distribution of low- and moderate-income housing within metropolitan areas has been another flash point in modern American society. The usual avenues toward "fair share" distribution of low-end housing have involved direct political negotiation among local governments (as pioneered by Ohio's Miami Valley Regional Planning Commission) or court orders to local jurisdictions (as with New Jersey's Mount Laurel decisions). Planning to meet Oregon's Goal 10 makes the fair share principle in the Portland area politically acceptable by requiring that every jurisdiction zone at least half of its vacant residential land for attached single-family or multi-family housing. In 1978-82 LCDC blocked efforts by several Portland suburbs to ignore or undercut the requirement. Since then it has been accepted that Oregon jurisdictions may not use zoning as a tool of social exclusion.

The Oregon planning program also defines metropolitan housing goals in terms of minimum rather than maximum densities. Traditional zoning in the United States establishes maximum numbers of housing units per acre by setting minimum lot sizes. The Oregon system begins to reverse the situation for the Portland area. The Metropolitan Housing Rule requires minimum allowable densities of ten housing units per net buildable acre in areas that contain most of the regional population. Housing planning therefore becomes a means for reinforcing the effectiveness of the UGB.

Another area of innovation is planning for conservation of offshore ocean resources. Goal 19 mandates the conservation of "the long-term values, benefits, and natural resources of the ocean both within the state and beyond." Oregon's Ocean Resources Management Plan, developed by a special task force with DLCD staff support and adopted by LCDC in 1990, defines a primary area of interest as the state's "territorial sea"— the ocean and ocean bottom extending 3 miles offshore.[3] It gives specific content to Goal 19 by defining ocean resources to include marine fisheries, marine birds and mammals, intertidal plants and animals, air and water quality, recreational and cultural resources, oil, gas, and minerals. In 1992-93, DLCD continued to provide staff for a new state Ocean Policy Advisory Council and for development of a more specific plan to define protections for nonrenewable resources and acceptable uses of renewable resources.[4] Originally triggered in the 1980s by concerns over offshore oil and gas exploration and mineral harvesting, the Ocean Resources Management Plan has attracted national attention for going beyond standard coastal zone planning to focus on tidelands and open ocean rather than a more narrow focus on developable coastal lands.

Ironically, despite this technical success and national reputation, the Oregon program continues to be vulnerable. Professional planners are frustrated, the public is complacent, and legislative support has eroded.

Professional planners, who should be among the strongest spokespersons on behalf of the system, are withholding their support.[5] Some are stymied because they are answerable to their employers who are in direct opposition to state planning directives. The greatest erosion of support has been among county planners who are dissatisfied with the compromises LCDC made in 1992 on the issue of small-scale resource lands. In releasing marginal land for limited development, LCDC imposed stricter requirements on more productive resource lands rather than admitting that some lands had been inappropriately classified when the original resource protection policies were developed.

Counties feel that they are not being heard by LCDC. On the other hand, LCDC feels that counties are being given ample opportunity for input. The Commission can point to innumerable public hearings, meetings, and opportunities for written comment. The counties can point to mandatory studies required of their understaffed departments which comply with strict deadlines only to discover that the work has to be redone six months later because of new LCDC-imposed rules. While no one has evaluated the legitimacy of these perceptions, they frame the debate that will ultimately determine the fate of the system.

And where is the public in this debate? There continues to be strong support for planning. For example, a coastal area planning director has seen an increasing acceptance of the importance and role of planning over the past ten years. He and his staff find that they need to spend less time in justifying their work than they did a decade ago. It is possible that this public support would enable the system to survive yet another referendum challenge. But this support may not translate into political influence because it tends to be passive. An entire generation of Oregonians have come of age since the creative/inclusive planning politics of the 1970s (passing SB 100, writing state goals, and preparing comprehensive plans). This new generation has experienced the Oregon system as a set of regulations rather than an envisioning process. They may be less committed to the system because they do not appreciate what it replaced and fail to see it as a system that they can use to actively protect the state's quality of life.

Afterword

Meanwhile, the people who are going to battle are the ones who are infuriated by the perceived injustices of the system—by the stories of a retired couple who are not allowed to build on resource land they bought thirty years ago or the small developer who was given the run-around and subsequently went bankrupt. When anecdotes like these make headlines and are not counterbalanced by testimonials, research, and information that can set individual experiences in a broader context, then the impression is conveyed that the system has serious shortcomings.

A significant source of frustration is grounded in a discrepancy between the intended type of relationships between governments and what has actually emerged. To be specific, the design for the Oregon planning system was envisioned as a partnership between the state in setting policy directives and local government in implementation. This was done in recognition of the very real difficulties faced by local politicians and professionals in regulating their constituency. Requirements imposed by a higher level of government enables those people on the front line to say, "the state makes me do this." And while many local officials chaff at the idea of the state taking away their decision-making options, others welcome the state's role as giving them the leverage to enforce basic planning principles. It is easy, however, to move from the notion of partnership to one of distrust of local actions by the state, particularly when the balance of power is uneven. Distrust leads to less flexibility, excessive regulation, underfunded mandates, and a breakdown in communication.

Festering controversies over the Oregon planning system came to a boil in the 1993 legislature. The conservative political agenda, which has gained influence nationally, had made inroads at the state and local level. Legislators were elected with a commitment to dismantling DLCD and, for the first time in history, supporters of the planning system could not account for a clear majority in either house. At the start of the session, Speaker of the House Larry Campbell announced that DLCD would lose its funding unless measures overturning LCDC's small-scale resource rules were adopted. Governor Barbara Roberts threatened to veto any effort to weaken the system and there were some strategists who envisioned a worst-case scenario of an intact system with no state agency to administer it.

Ultimately, in the waning days of the record-length session, a compromise emerged in the form of HB 3661, which maintained strong protection of high-value farm land and gave development rights to less productive land that had been purchased before 1985. For those who had feared the worst, the outcome was surprisingly positive. The legislation made certain aspects of the state's resource protection policies easier to implement. Clear definitions and standards were spelled out, replacing a negotiation process which had promised to place a significant strain on DLCD staff. Increased protection was provided by removing the possibility of any more counties opting for the more permissive marginal lands policy alternatives. Right-to-farm provisions were directed toward ensuring that rural development did not impose a hardship on commercial agriculture operations.

DLCD was subsequently funded at a level roughly equivalent to the governor's original request.

Some participants in the program are optimistic that the worst is behind them. They hope that passage of HB 3661 will translate into long-term support. However, while there is widespread agreement that HB 3661 is better than had been anticipated, it is not yet clear that the right provisions were put into the legislation. The answer will become apparent during the rule-making process and subsequent implementation. Deschutes County planners, for example, have determined that 70 percent of their parcels have changed hands since 1985, lessening the relevance of the emphasis on lots of record. They are more concerned with provisions that will allow marginal lands to be considered separately from a larger farm unit. They also wonder what the local reaction will be if they are required to make significant changes in the structure of their agricultural zoning, which underwent a wholesale review and modification in 1992.

Recommendations

The Oregon planning system merits continued support. It is important to acknowledge that problems exist. But rather than wishing away the system because of these problems, Oregonians need to recognize that the system provides the means for making improvements in the way we plan and regulate. Anyone who has been involved in local planning and development in states where "anything goes" will recognize that Oregon's land use planning program as a system is an extraordinary asset that makes it possible to improve the way we do business.

It is imperative that the system maintain a meaningful partnership among levels of government. The heart of the program is identifying state policy concerns and developing appropriate guidelines and regulations. Where there is no legitimate state-level concern, local governments should be given more flexibility. They might be given the option of choosing from a variety of ways to achieve state goals. This flexibility might involve explicit recognition of differences among the coastal counties, Willamette Valley, and eastern Oregon.

The legislature should provide a budget that gives local governments the resources to comply with state mandates. It should also enable DLCD to provide technical assistance to local governments.

LCDC should better utilize and empower its standing statewide advisory committee on citizen involvement which to date has not been used effectively. This committee could be charged with examining the complaints that LCDC and DLCD are working in isolation and develop recommendations for addressing this problem.

The Oregon system has done an exceptional job in extending the time frame used in planning. Twenty years, which was the basis for defining UGBs, is the norm for American planning practice. But as communities come up against the end of the first twenty years, it becomes readily apparent that two decades is not long enough. The Metropolitan Service District (Metro) has now undertaken the 2040 project to set goals for the next fifty years; the urban reserve rule concept is also based on the same time frame. Meanwhile a growing number of Oregon planners are talking in terms of one hundred years, which encourages a perspective that builds on the notion of sustainable development and leads to a balance between population and the environment that can be maintained in perpetuity.

The Oregon system will lose its base of support unless more attention is given to communicating in plain English what is being accomplished. Oregonians need to understand that their system is working. They also need to understand why planning is important and what could happen in its absence. Planning is not an instinct; it is a process that is learned over time. It builds on a set of values that must be reaffirmed as the actors change.

The Oregon planning system might benefit from a university "home," a place where academics take responsibility for evaluation, research, and reflection. This would enable both state and local policy makers to gain a better understanding of ways in which the system can be improved, thus avoiding the more serious problems that have been so damaging. The university could be the advocate for the system as a whole, a perspective that is often lost as policy makers and special interest groups haggle over more narrowly defined concerns.

The university can also be a source of education and training. It can give future professionals the specific skills needed to work within the system and help citizens become more effective. Education needs to include Oregon's children—the land users, policy makers, and planners of the future. To this end, primary and secondary education curricula should include an introduction to Oregon's planning system.

Ultimately the survival of the Oregon planning program will depend on the extent to which Oregonians can create and maintain a culture of learning. This involves developing an understanding of the strengths and weaknesses of the system and creating the means by which new ideas can be articulated, assessed, and where appropriate, adopted. The process of adaptation will be ongoing and at times will be painful. Patience, sensitivity, and compromise are essential. At the same time it is important not to lose sight of the larger goals of resource management and the creation of vibrant, healthy communities. The Oregon system has the potential for being the most significant legacy that today's citizens can bequeath to the future. It is through planning that we can ensure that today's decisions can maximize tomorrow's alternatives.

Notes

1. The general planning policies for the Cape Cod Seashore were developed by a Cape Cod Commission with membership drawn from local, state, and federal governments.
2. Urban reserves are lands outside UGBs that are targeted for eventual UGB expansion. A state planning policy promulgated in 1992 requires special planning for such lands to assure their eventual suitability for urbanization (e.g., by preventing their fragmentation into multi-acre residential parcels).
3. The 1953 federal Submerged Lands Act established coastal states' ownership of the sea bottom within 3 miles of their coast.
4. Ocean planning, of course, is hampered by the paucity of baseline inventory and trend data for ocean resources.
5. A significant exception is the formation of the Legislative Policy Action Committee (LPAC) by the Oregon chapter of the American Planning Association in late 1992. The LPAC is specifically oriented toward providing support for the Oregon planning system.

Oregon's Statewide Planning Goals

Goal 1: Citizen Involvement
To develop a citizen involvement program that insures the opportunity for citizens to be involved in all phases of the planning process.

Goal 2: Land Use Planning
To establish a land use planning process and policy framework as a basis for all decisions and actions related to use of land and to assure an adequate factual base for such decisions and actions.

Goal 3: Agricultural Land
To preserve and maintain agricultural lands.

Agricultural lands shall be preserved and maintained for farm use, consistent with existing and future needs for agricultural products, forest and open space and with the state's agricultural land use policy expressed in ORS 215.243. Oregon's agricultural diversity is reflected in farms of different sizes and varying production levels. To reflect this agricultural diversity, counties may adopt different types of comprehensive plan designations to protect agricultural lands.

Counties, in cooperation with LCDC, the Departments of Agriculture and Forestry and farm and forestry experts, may identify and map small-scale resource lands. LCDC, in cooperation with counties, the Department of Agriculture, farmers and agricultural experts, may identify high-value farmland. Agricultural lands not identified as high-value farmland or small-scale resource lands shall be identified as important farmlands. High-value and important farmland shall be preserved and maintained for commercial farm use.*

---

*This paragraph will be substantially altered to comply with the actions of the 1993 legislature. Changes are to be adopted by March 1, 1994.

Goal 4: Forest Lands

To conserve forest lands by maintaining the forest land base and to protect the state's forest economy by making possible economically efficient forest practices that assure the continuous growing and harvesting of forest tree species as the leading use on forest land consistent with sound management of soil, air, water, and fish and wildlife resources and to provide for recreational opportunities and agriculture.

Goal 5: Open Space, Scenic and Historic Areas, and Natural Resources

To conserve open space and protect natural and scenic resources.

Programs shall be provided that will (1) insure open space, (2) protect scenic and historic areas and natural resources for future generations, and (3) promote healthy and visually attractive environments in harmony with the natural landscape character.

Goal 6: Air, Water, and Land Resources Quality

To maintain and improve the quality of the air, water and land resources of the state.

All waste and process discharges from future development, when combined with such discharges from existing developments shall not threaten to violate, or violate applicable state or federal environmental quality statutes, rules and standards. With respect to the air, water and land resources of the applicable air sheds and river basins described or included in state environmental quality statutes, rules, standards and implementation plans, such discharges shall not (1) exceed the carrying capacity of such resources, considering long range needs; (2) degrade such resources; or (3) threaten the availability of such resources.

Goal 7: Areas Subject to Natural Hazards and Disasters

To protect life and property from natural disasters and hazards.

Developments subject to damage or that could result in loss of life shall not be planned nor located in known areas of natural disasters and hazards without appropriate safeguards. Plans shall be based on an inventory of known areas of natural disaster and hazards.

Goal 8: Recreational Needs

To satisfy the recreational needs of the citizens of the state and visitors and, where appropriate, to provide for the siting of necessary recreational facilities including destination resorts.

Goal 9: Economic Development

To provide adequate opportunities throughout the state for a variety of economic activities vital to the health, welfare, and prosperity of Oregon's citizens.

Goal 10: Housing

To provide for the housing needs of citizens of the state.

Buildable lands for residential use shall be inventoried and plans shall encourage the availability of adequate numbers of needed housing units at price ranges and rent levels which are commensurate with the financial capabilities of Oregon households and allow for flexibility of housing location, type and density.

Goal 11: Public Facilities and Services

To plan and develop a timely, orderly and efficient arrangement of public facilities and services to serve as a framework for urban and rural developments.

Goal 12: Transportation

To provide and encourage a safe, convenient and economic transportation system.

A transportation plan shall (1) consider all modes of transportation including mass transit, air, water, pipeline, rail, highway, bicycle and pedestrian; (2) be based upon an inventory of local, regional and state transportation needs; (3) consider the differences in social consequences that would result from utilizing differing combinations of transportation modes; (4) avoid principle reliance upon any one mode of transportation; (5) minimize adverse social, economic and environmental costs; (6) conserve energy; (7) meet the needs of the transportation disadvantaged by improving transportation services; (8) facilitate the flow of goods and services so as to strengthen the local and regional economy; and (9) conform with local and regional comprehensive plans.

Goal 13: Energy Conservation

To conserve energy.

Land and uses developed on the land shall be managed so as to maximize the conservation of all forms of energy, based upon sound economic principles.

Goal 14: Urbanization

To provide for an orderly and efficient transition from rural to urban land use.

Urban growth boundaries shall be established to identify and separate urbanizable land from rural land. Establishment and change of the boundaries shall be based upon considerations of the following factors: (1) demonstrated need to accommodate long-range urban population growth requirements consistent with LCDC goals; (2) need for housing, employment opportunities, and livability; (3) orderly and economic provision for public facilities and services; (4) maximum efficiency of land uses within and on the fringe of the existing urban areas; (5) environmental, energy, economic and social consequences; (6) retention of agricultural land with Class I being the highest priority for retention and Class VI the lowest priority; and (7) compatibility of the proposed urban uses with nearby agricultural activities.

Goal 15: Willamette River Greenway

To protect, conserve, enhance and maintain the natural, scenic, historical, agricultural, economic and recreational qualities of lands along the Willamette River as the Willamette River Greenway.

Goal 16: Estuarine Resources

To recognize and protect the unique environmental, economic and social values of each estuary and associated wetlands; and to protect, maintain, where appropriate develop, and where appropriate restore the long-term environmental, economic, and social values, diversity and benefits of Oregon's estuaries.

Goal 17: Coastal Shorelands

To conserve, protect, where appropriate develop and where appropriate restore the resources and benefits of all coastal shorelands, recognizing their value for protection and maintenance of water quality, fish and wildlife habitat, water-dependent uses, economic resources and recreation and aesthetics. The management of these shoreland areas shall be compatible with the characteristics of the adjacent coastal waters; and to reduce the hazard to human life and property, and the adverse effects upon water quality and fish and wildlife habitat, resulting from the use and enjoyment of Oregon's coastal shorelands.

Goal 18: Beaches and Dunes

To conserve, protect, where appropriate develop, and where appropriate restore the resources and benefits of coastal beach and dune areas; and to reduce the hazard to human life and property from natural or man-induced actions associated with these areas.

Goal 19: Ocean Resources

To conserve the long-term values, benefits, and natural resources of the nearshore ocean and the continental shelf.

All local, state, and federal plans, policies, projects, and activities which affect the territorial sea shall be developed, managed and conducted to maintain and where appropriate, enhance and restore, the long-term benefits derived from the nearshore oceanic resources of Oregon. Since renewable ocean resources and uses, such as food production, water quality, navigation, recreation, and aesthetic enjoyment, will provide greater long-term benefits than will nonrenewable resources, such plans and activities shall give clear priority to the proper management and protection of renewable resources.

Adoption

Goals 1-14	December 27, 1974
Goal 15	December 6, 1975
Goals 16-19	December 18, 1976

Amendment:

Goals 2-4	December 30, 1983
Goal 8	October 19, 1984
Goals 16-19	October 19, 1984
Goals 1, 2, 3, 5, 8-11, 14, 15	February 17, 1988
Goal 4	January 25, 1990
Goals 3-4	August 7, 1992

Annotated Bibliography

Journal articles

Abbott, Carl, and Deborah Howe. "The Politics of Land-Use Law in Oregon: Senate Bill 100, Twenty Years After." *Oregon Historical Quarterly* Vol. 94 (1). Spring 1993, pp. 5-35.

Introduction and transcription of interview with four key actors in the formation and development of the Oregon land use program. Interviewees are Hector Macpherson, Ted Hallock, Stafford Hansell, and Henry Richmond.

Bollens, Scott A. "State Growth Management: Intergovernmental Frameworks and Policy Objectives." *Journal of the American Planning Association* Vol. 58 (4). Autumn 1992, pp. 454-466.

In a comparative analysis that includes Oregon, the author finds that state growth management programs have shifted over the years from preemptive regulation by states to cooperative state/local planning.

Buckland, Jeffrey G. "Growth Management: Two County Approaches in the Pacific Northwest." *Journal of Soil and Water Conservation* Vol. 41(6). November/December 1986, pp. 383-385.

The author compares growth management in Multnomah County, Oregon, and King County, Washington. Concludes that there is a need for better information to assess how the policies are working. Both counties are just beginning to evaluate their efforts.

Coughlin, Robert E., and John C. Keene. "The Protection of Farmland: An Analysis of Various State and Local Approaches." *Land Use Law and Zoning Digest* Vol 33(6). June 1981, pp. 5-11.

Reviews and analyzes the actions that states (including Oregon) have taken to slow the pace at which farmland is being converted to nonfarm uses.

Daniels, Thomas L., and Arthur C. Nelson. "Is Oregon's Farmland Preservation Program Working?" *Journal of the American Planning Association* Vol. 52(1). Winter 1986, pp. 22-32.

The author states that Oregon's farm land preservation program appears to have been successful in keeping the state's farmland from being converted to nonfarm uses. However the proliferation of small hobby farms raises concerns about the future viability of commercial farming operations which must compete for the same farm land.

Davis, Gordon. "Special Area Management: Resolving Conflicts in the Coastal Zone." *Environmental Comment* Oct. 1980, pp. 4-7.

This case study of special area management in Coos Bay, Oregon, describes some of the principles and techniques required to carry out a special management program. It provides an example of the resolution of the question of using a large freshwater marsh for future industrial lands.

Furuseth, Owen J. "The Oregon Agricultural Protection Program: a Review and Assessment." *Natural Resources Journal* Vol. 20(3). July 1980, pp. 603-614.

The article examines the components of the Oregon farm land protection program, the relative success of the program, and its unique characteristics.

Furuseth, Owen J. "Update on Oregon's Agricultural Protection Program: A Land Use Perspective." *Natural Resources Journal* Vol. 21(1). January 1981, pp. 57-70.

Using the 1978 Census of Agriculture, the author examines the changes in agricultural land use in Oregon since the land use program was adopted.

Fussner, Sarah Elizabeth, and Wiley, William S. "Oregon's New State Land Use Planning Act—Two Views." *Oregon Law Review* Vol. 54(20). 1975, pp. 203-223.

Two contrasting views of the Oregon Land Use Act are presented. The first discusses the Land Conservation and Development Commission's power to grant permits and review plans and suggests that state-level planning should be limited. The second compares the Act to the American Law Institute's Model Land Development Code and the defeated Land Use Policy and Planning Assistance Act 1972.

Gale, Dennis E. "Eight State-Sponsored Growth Management Programs: A Comparative Analysis." *Journal of the American Planning Association* Vol. 58 (6). Autumn 1992, pp. 425-39.

This systematic presentation of program features and provisions compares Oregon with seven other states. The author classifies Oregon along with Florida, Maine, and Rhode Island as a "state dominated" program.

Gangle, Sandra Smith. "LCDC Goal 10: Oregon's Solution to Exclusionary Zoning." *Willamette Law Review* Vol. 16(3). Summer 1980, pp. 873-889.

The Land Conservation and Development Commission's decisions which interpret Goal 10: Housing are analyzed. From these decisions, a housing policy has evolved which is a unique approach to the problem of exclusionary zoning.

Gordon, S. C. "Urban Growth Management, Oregon Style." *Public Management* Vol. 70(8). August 1988. pp. 9-11.

An examination of how Oregon's land use program has managed urban growth, with Eugene-Springfield as an example.

"Growth Management Workshop: The Compelling Forces Behind Six Municipal Efforts to Control Urban Development." *Environmental Comment* June 1979, pp. 5-16.

Issue reviews presentations made at a May 1979 Urban Land Institute workshop on growth management programs. The presentations compared the growth management plans and results of six urban communities including Salem, Oregon.

Gustafson, Greg, Thomas L. Daniels, and Rosalyn P. Shirack. "The Oregon Land Use Act: Implications for Farmland and Open Space Protection." *Journal of the American Planning Association* Vol 48(3). Summer 1982, pp. 365-373.

Oregon's combination of state mandated, locally implemented urban growth boundary designations and exclusive farm use zoning represents a unique case in farm land protection policy. The performance of the program is evaluated and economic trade-offs in the selection of minimum lot size standards are discussed.

Honey, Keith M. "Land Use Planning." *Practicing Planner* Vol 8(2). June 1978, pp. 15-17, 20.

Extensive conversations with professional planners in Colorado, Washington, Oregon, and California are the basis for a summary of the most prevalent and critical problems associated with land use planning in those states.

Knaap, Gerrit J. "Social Organization, Profit Cycles and Statewide Land Use Controls: Welcome to Oregon—Enjoy Your Visit." *Journal of Applied Behavioral Science* Vol. 23(3). 1987, pp. 371-385.

Offers a political and economic analysis of statewide land use controls, focusing on the land use program in Oregon. Using empirical evidence, based on the state's population characteristics, industries and elections featuring referenda to repeal the land use statutes, the author suggests that statewide land use controls in Oregon are supported by identifiable private interests.

Knaap, Gerrit J. "The Price Effects of Urban Growth Boundaries in Metropolitan Portland, Oregon." *Land Economics* Vol 61(1). February 1985, pp. 26-35.

Using cross-section data, the study measures the effects of urban growth boundaries on vacant single-family land values in metropolitan Portland, Oregon. The boundaries were found to have a significant influence. Nonurban land inside the UGB's will be prepared for conversion to urban use, while nonurban land outside UGBs will remain free of speculative influences.

Knaap, Gerrit J. "State Land Policy and Exclusionary Zoning: Evidence from Oregon." *Journal of Planning Education and Research* Vol. 9 (2). 1990, pp. 39-46.

Analysis of the ways in which state planning in Oregon fosters a limited form of inclusionary zoning. The study suggests that differences in political environment between state and local levels offer the potential to overcome a major obstacle to providing low-income housing.

Liberty, Robert L. "Oregon's Comprehensive Growth Management Program: An Implementation Review and Lessons for Other States." *Environmental Law Reporter* 22 (6). 1992, pp. 10367-10391.

A detailed analysis of the legal aspects of implementation of the Oregon system. The discussion includes an analysis of aspects of Oregon's system that might be considered by other states.

Medler, Jerry, and Alvin Mushkatel. "Urban-Rural Class Conflict in Oregon Land Use Planning." *Western Political Quarterly* Vol. 32(3). September 1979, pp. 338-349.

This article looks at citizen support for and opposition to Oregon's land use regulations as reflected in the 1976 referendum to repeal the statutes. Concludes that the better off counties and cities favor land use planning while the less well off reject it.

Morgan, Terry D. "Exclusionary Zoning: Remedies Under Oregon's Land Use Planning Program." *Environmental Law* Vol 14(4). Summer 1984, pp. 779-830.

Using a recent case, the author explores the methods available under the land use program to curb exclusionary zoning. Part of the proceedings of a symposium on Oregon land use held February 17-18, 1984 at the Northwest School of Law.

Morgan, Terry D., and Shonkwiler, John W. "Urban Development and Statewide Planning: Challenge of the 1980s." *Oregon Law Review* Vol. 61(3). 1982, pp. 351-394.

The article examines the current land use planning laws and contrasts the problems of centralization with the advantages of decentralization. Suggestions for the transition to decentralized urban planning are offered.

Muzzalli, William B. "The Future of Oregon's Land Use Appeals Process: Sunset on LUBA." *Willamette Law Review* Vol. 19(1). Winter 1983, pp. 109-137.

Discusses the Land Use Board of Appeals' legislative history and operation. Also examines proposals the 1983 Oregon legislature would be considering regarding the future of the board.

Nelson, Arthur C. "Reader Response." *Natural Resource Journal* Vol. 23(1). January 1983, pp. 1-3.

Reponse to Furnseth's articles on Oregon's agricultural land use planning program. Says that Furuseth is premature to suggest that program is any more effective at improving agricultural trends than the lack of such a program.

Nelson, Arthur C. "Demand, Segmentation, and Timing Effects of an Urban Containment Program on Urban Fringe Land Values." *Urban Studies* Vol. 22 (5). October 1985, pp. 439-443.

Offers empirical evidence on the market effects of an urban containment program on urban fringe land values in the Salem area. The evidence suggests that the land market internalizes the supply restriction effect within four years.

Nelson, Arthur C. "Using Land Markets to Evaluate Urban Containment Programs." *Journal of the American Planning Association* Spring 1986, pp. 156-171.

The author develops a theory of how urban containment programs should influence the regional land market, and then applies the theory to a case study. Results are threefold. First, the urban containment program employed by Salem, Oregon, separates the regional land market into urban and rural components. Second, by making greenbelts out of privately held farm land, the program prevents speculation on farm land in the regional land market. Third, greenbelts add an amenity value to urban land near them.

Nelson, Arthur C. "An Empirical Note on how Regional Urban Containment Policy Influences an Interaction between Greenbelt and Exurban Land Markets." *Journal of the American Planning Association* Vol. 54 (2): 1988, pp. 178-184.

Uses data from Washington County to model land values, concluding that exurban land and farm land constitute separate markets. Although exurban land benefits from proximity to greenbelt amenities, the two markets can coexist.

Nelson, Arthur C. "Preserving Prime Farmland in the Face of Urbanization: Lessons from Oregon." *Journal of the American Planning Association* Vol. 58 (4). Autumn 1992, pp. 467-488.

The author evaluates the sucesses and mistakes of Oregon's system of farm land protection through comprehensive planning, farm tax deferral, and creation of a dual land market inside and outside urban growth boundaries.

Niemi, Ernest G. "Oregon's Land Use Program and Industrial Development: How Does the Program Affect Oregon's Economy?" *Environmental Law* Vol 14(4). Summer 1984, pp. 707-712.

> The author concludes that the effect of Oregon's land use program on the state's economy should be measured by the cost of doing business in Oregon, rather than by the number of acres designated for industrial development. Part of the proceedings of a symposium on Oregon land use held February 17-18, 1984, at the Northwest School of Law.

Pease, James R. "Land Use Designation in Rural Areas: An Oregon Case Study." *Journal of Soil and Water Conservation* Vol. 45 (5). September-October 1990, pp. 524-528.

> Reviews initial efforts to define primary and secondary rural lands. Addresses policy issues and technical criteria and identifies problems with citizen participation, interest group involvement, and utilization of technical information in policy making.

Porter, Douglas R. "LCDC, UGB, and EFU's: Oregon's Pioneering Land Use Program." *Urban Land* Vol. 42(5). May 1983, pp. 34-35.

> A brief review of the history, accomplishments and failures of Oregon's form of development regulation, written for a readership of persons interested in the land conversion and real estate development process.

Richmond, Henry R. "Unique Public Interest Law Group Supports Oregon Land Use Program." *Land Use and Zoning Digest* Vol. 31(10). October 1979, pp. 4-8.

> The executive director of 1000 Friends of Oregon describes the role played by the organization in executing Oregon's land use program.

Richmond, Henry R. "Does Oregon's Land Use Program Provide Enough Desirable Land to Attract Industry to Oregon?" *Environmental Law* Vol. 14(4). Summer 1984, pp. 693-706.

> The question of whether the Oregon land use program hinders industrial development is discussed. The author concludes that the program has helped, rather than hindered, industrial development. Part of the proceedings of a symposium on Oregon land use held February 17-18, 1984, at the Northwest School of Law.

Ross, James F. "Land Use Planning and Coastal Zone Management—The Oregon Story." *Environmental Law* Vol. 5 (3). Spring 1975, pp. 661-673.

> The relationship between coastal zone management and land use planning within Oregon is described, as well as the role of the public in developing state land and water use planning efforts.

Schell, Steven R. "Living with the Legacy of the 1970's: Federal/State Coordination in the Coastal Zone." *Environmental Law* Vol 14(4). Summer 1984, pp. 751-778.

The conflicts caused by overlapping state and federal coastal zone jurisdiction are examined. The author concludes with five maxims for dealing with the legal, economic and environmental problems of the coastal zone. Part of the proceedings of a symposium on Oregon land use held February 17-18, 1984, at the Northwest School of Law.

Shurts, John. "Goal Four and Nonforest Uses on Forest Lands." *Environmental Law* Vol. 19(1). Fall 1988, pp. 59-91.

The author states that the Land Conservation and Development Commission has not followed a consistent aproach for establishing nonforest uses on forest lands. He concludes that the commission should clarify the approach it takes toward nonforest use policy by articulating a comprehensive policy together with interpretive rules.

Sullivan, Edward J. "Spectre at the Celebration: Will San Diego Gas Hinder Oregon's Ten-Year-Old Land Use Program?" *Environmental Law* Vol 14(4). Summer 1984, pp. 661-692.

This article describes three approaches to constitutional interpretation in the context of fifth and fourteenth amendment takings analysis. It also attempts to predict the effect of *San Diego Gas and Electric v. San Diego* on Oregon's land use program. Part of the proceedings of a symposium on Oregon land use held February 17-18, 1984, at the Northwest School of Law.

Sullivan, Edward J., Bernard H. Siegan, and Norman Williams, Jr. "Panel Discussion: the Oregon Example: A Prospect for the Nation." *Environmental Law* Vol 14(4). Summer 1984, pp. 843-862.

Panelists discuss the successes and failures of Oregon's land use program and compare it to programs in other states. Part of the proceedings of a symposium on Oregon land use held February 17-18, 1984, at the Northwest School of Law.

Sullivan, Edward J. "Land Use and the Oregon Supreme Court: A Recent Retrospective." *Willamette Law Review*. Vol 25(2). Spring 1989. P. 259-292.

This article uses the opinions of two members of the Oregon Supreme Court to sample how the state's land use program has fared in that court.

Thatcher, Terrence L., and Duhnkrack, Nancy. "Goal Five: The Orphan Child of Oregon Land Use Planning." *Environmental Law* Vol 14(4). Summer 1984, pp. 713-750.

This article traces the inadequacy of implementing Goal Five—conserving open space and scenic resources—and criticizes the Goal Five Administrative Rule. Part of the proceedings of a symposium on Oregon land use held February 17-18, 1984, at the Northwest School of Law.

Newsletters and Periodicals

Bureau of Governmental Research and Service, University of Oregon. *Index to Recent Oregon Land Use Decisions.* Vols. 1-3 (1985-87).

Topical guide to land use decisions by Oregon judicial system.

Department of Land Conservation and Development. *Oregon Lands.* Newsletter of the Department of Land Conservation and Development. Salem, Oregon. Published from 1978 to 1980. Monthly/Quarterly. 6-12 pages.

Provides information on LCDC's actions, policies, procedures, and impacts as well as other aspects of Oregon's planning program.

Department of Land Conservation and Development. *Oregon Planning News.* Published since 1984.

Newsletter describing DLCD activities and programs.

1000 Friends of Oregon. *1000 Friends of Oregon Newsletter.* Portland, Oregon. Published from 1975 to present. Monthly/Quarterly. 4-10 pages.

News, editorials, and articles about the Oregon's land use program from 1000 Friends of Oregon, a nonprofit membership organization dedicated to protecting Oregon's land use laws.

1000 Friends of Oregon. *Landmark.* 1000 Friends of Oregon. Portland, Oregon. Published from 1983 to present. 24-36 pages.

Quarterly journal of 1000 Friends of Oregon, a nonprofit membership organization dedicated to protecting Oregon's land use laws. The articles review land use court decisions and discuss current planning issues.

Oregon Land Use Board of Appeals. *Reports of Cases Decided in the Oregon Land Use Board of Appeals.* Salem: The Board. Vol. 14- (1985-).

The standard legal source for text of LUBA decisions. Continues the following series:

Oregon Land Conservation and Development Commission Decisions. Seattle: Butterworth. Vols. 1-3 (1973-80).

Oregon Land Use Board of Appeals Decisions. Seattle: Butterworth. Vols. 1-13 (1980-85).

Books

DeGrove, John. 1984. *Land, Growth and Politics.* Chicago: American Planning Association Planners Press. 454 pages.

The now classic study of land use planning and growth management responses to rapid urbanization, with considerable attention to the Oregon system. The book has had substantial influence on the "second generation" of state growth management programs since the mid-1980s.

DeGrove, John. 1992. *The New Frontier for Land Policy: Planning and Growth Management in the States.* Cambridge: Lincoln Institute of Land Policy.

A comprehensive overview of recent state-level land use planning efforts around the United States, with discussion of the influence of Oregon.

Knaap, Geritt, and A. C. Nelson. 1992. *The Regulated Landscape: Lessons on State Land Use Planning from Oregon.* Cambridge: The Lincoln Institute of Land Policy.

A detailed case study of the Oregon land use planning system after two decades. The discussion includes attention to effects on land values, farming, economic development, and the wood products industry.

Leonard, H. Jeffery. 1983. *Managing Oregon's Growth.* Washington: The Conservation Foundation. 160 pages.

Examines the experiences of the state government and local governments as they implemented the statewide land use planning program. Specifically looks at the politics of local planning, efforts to protect rural land and manage urban growth, and the future of Oregon's program.

Little, Charles E. 1974. *The New Oregon Trail.* Washington: The Conservation Foundation. 37 pages.

An account of the development and passage of state land use legislation in Oregon. Includes an interview with Governor Tom McCall.

Oregon Land Use Board of Appeals Handbook. 1981. Seattle: Butterworth Legal Publications.

Contains statewide planning goals and guidelines, Oregon Administrative Rules relating to LCDC, and LUBA rules of procedure.

Rohse, Mitchell. 1988. *Land-Use Planning in Oregon.* Corvallis: Oregon State University Press.

A concise guide to the Oregon system for citizens, planning commissioners, and students. Includes a brief outline of the system and definitions of key terms.

Stein, Jay M., ed. 1993. *Growth Management: The Planning Challenge of the 1990s.* Newbury Park, CA: Sage Publications.

Includes a chapter by Deborah Howe on "Growth Management in Oregon" that assesses the wide range of efforts at resource protection and urban growth management and argues the continued viability of Oregon's program.

Steiner, Frederick, and Theilacker, John, eds. 1984. *Protecting Farmlands.* Westport, CT: AVI Publishing Co.

Includes a chapter by Ronald Eber on "Oregon's Agricultural Land Protection Program."

Documents, Theses, and Reports

Bureau of Governmental Research and Service. *Oregon Land Use Planning: Local Planning Digest.* Eugene: University of Oregon. November 1984.

Provides an overview of Oregon's land use planning and development system, prepared especially for city and county elected officials and planning commission members. It is a summary of the lengthier and more technical *Guide to Local Planning and Development*, also published by the Bureau of Governmental Research and Service as part of its Land Use Training Materials Package.

Bureau of Governmental Research and Service. *Oregon Land Use Planning: Case Studies.* Eugene: University of Oregon. November 1984.

A collection of case studies illustrating several types of land use actions and procedures involved in the implementation of comprehensive plans by various local governments in Oregon. It is designed to help local officials and others involved in the land use process better understand their own systems and practices.

Bureau of Governmental Research and Service. *Land Use Procedures and Practices in Oregon.* Eugene: University of Oregon. January 1985.

A manual describing local government land use decision-making processes. Presents some examples of forms and documents used in land use management as examples of how some jurisidictions handle various land use matters. Intended to assist the everyday work of practicing planners.

Bureau of Governmental Research and Service. *Local Planning Digest.* Eugene: University of Oregon. 1986.

A summary of Oregon planning laws aimed at local planning officials and practitioners.

Bibliography

Chrisman, Janis, and Judity Armatta. *Coastal Shorelands and the Need for Senate Bill 100.* Oregon Student Public Interest Research Group. October 1976.

A review of the legal and regulatory authority of 14 agencies with major responsibilities on the Oregon coast. Analyzes two laws with particular applicability to coastal regions: the federal Coastal Zone Management Act and the state Senate Bill 100. Recommendations are made on the draft coastal shorelands goals.

DeGrove, John, and Nancy Stroud. *Oregon's State Urban Strategy.* Washington: Government Printing Office. 1980.

Historical and analytical review of the the urban growth management aspects of the state planning system. Compares the planning systems and associated political controversy in the states of California, Colorado, Florida, Hawaii, North Carolina, Oregon, and Vermont.

Governor's Task Force on Land Use. *Governor's Task Force on Land Use in Oregon: Report to Governor Vic Atiyeh.* Stafford Hansell, Chairman. September 1982.

Evaluation of the impacts of Oregon's land use planning program conducted for the governor. Areas of special interest include the length of time required to complete acknowledged plans, efficiencies and inefficiencies in the permit process, problems of plan implementation, and land use litigation.

Groll, Bruce J. *Oregon Comprehensive Land Use Planning.* Thesis. Willamette University. Atkinson Graduate School of Management. Center Paper 82-3. March 14, 1982.

A history and evaluation of land use planning in Oregon with recommendations for the future.

Halprin and Associates. *Willamette Valley: Choices for the Future.* Salem: State of Oregon. 1972.

This report by a California landscape architecture firm helped to shape the debate on Oregon land use by raising the possibility of urban sprawl throughout the Willamette Valley.

Joint Legislative Committee on Land Use. *Final Report of the Joint Legislative Committee on Land Use.* Chairman Representative Stephen Kafoury. November 1976. 109 pages.

A detailed chronicle of the first two years of Oregon's land conservation and development program. It also covers the committtee's findings of how the land use program relates to compensatory zoning and tax assessment.

Ketcham, Paul W., and Robert E. Stacy. *Analysis of Oregon's Forestry Program for Compliance with Statewide Planning Goals.* Portland: 1000 Friends of Oregon. 1985. 35 pages.

A report on research determining the level of fish and wildlife protection afforded by forest operations in Goal 5 and 17 resource areas.

Knaap, Gerrit J. *The Price Effects of an Urban Growth Boundary: A Test for the Effects of Timing.* Ph.D. dissertation, University of Oregon. 1982.

   An application of economic theory and measures to Oregon's chief growth management tool.

Land Conservation and Development Commission *The Oregon Ocean Plan.* Salem: Department of Land Conservation and Development. 1991.

   A pioneering plan for state involvement in management of ocean resources. The Ocean Plan is the state of Oregon's policy guide for management of state and federal waters off Oregon's coast.

Land Conservation and Development Commission. *Statewide Goals and Guidelines.* Salem: Department of Land Conservation and Development. 1976.

   The basic document that sets forth the goals of the Oregon planning system, including the complete text of all nineteen goals and guidelines. All local comprehensive plans must indicate the ways in which these goals have been utilized in setting local land use planning policies. Periodically reprinted.

Land Conservation and Development Commission. *Urban Growth Management Study: Summary Report.* Salem: Department of Land Conservation and Development. 1991.

   Summarizes detailed studies by independent consultants of three issues of urban growth management: (1) recent development inside urban growth boundaries; (2) development outside urban growth boundaries; (3) infrastructure funding. The complete study has eleven volumes including case studies of Portland, Bend, Medford, and Brookings.

Land Conservation and Development Commission. *1985-1987 Biennial Report to the Legislative Assembly of the State of Oregon.* Salem: Department of Land Conservation and Development. James F. Ross, Director. January 1987. 12 pages.

   This biennium marked the end of the first phase of the planning program and the commission's review and acknowledgment of local comprehensive plans. The report describes the state of the land use planning program at the end of that phase.

Miller, Tamara E. *The Two Oregons: Comparing Economic Conditions Between Rural and Urban Oregon.* Report to Joint Legislative Committee on Trade and Economic Development. 1990.

   Summary of statistical data that sets the context for state policy debates on rural growth and economic development.

Oregon Benchmarks: Setting Measurable Standards for Progress. Salem: Oregon Progress Board. January 1991.

A pioneering report setting measurable targets for state social, economic, and environmental policy, including land use, housing, and transportation goals.

Oregon State Bar, Committee on Continuing Legal Education. *Land Use.* 1982, with 1988 supplement.

Two volumes on statutory and case law of planning in Oregon. The thirty-six chapters cover both statewide goals and implementation of land use regulations.

Short, Sharyl Elaine. *County Responses to Goal 5 of LCDC Planning Goals and Guidelines.* M.A. thesis, Oregon State University. 1982.

This outlines some of the counties' responses to Goal 5. It discusses the attitude and training of county planners responsible for implementing the Goal 5 resources review, and concludes that the vagueness of goal requirements contribute to a conflict situation.

Zachary, Kathleen Joan. *Politics of Land Use: The Lengthy Saga of Senate Bill 100.* M.A. thesis, Portland State University. 1978.

A detailed legislative history of the development and passage of Senate Bill 100 which designated the state land use planning organizational structure.

Contributors

Carl Abbott is professor of Urban Studies and Planning at Portland State University, where he teaches courses on urban history and policy. He is the author of several books, including *Portland: Planning, Politics and Growth in a Twentieth Century City* (Lincoln: University of Nebraska Press, 1983); *The New Urban America: Growth and Politics in Sunbelt Cities* (Chapel Hill: University of North Carolina Press, 1987); and *The Metropolitan Frontier: Cities in the Modern American West* (Tucson: University of Arizona Press, 1993).

Sy Adler is professor of Urban Studies and Planning at Portland State University. His fields of interest include the implementation of planning in different political contexts and urban transportation policy. Articles on the latter subject have appeared in such journals as *Environment and Planning, Urban Affairs Quarterly*, and the *Journal of Policy History*.

John DeGrove is the director of the Joint Center for Environmental and Urban Problems of Florida Atlantic University and Florida International University. He is recognized as a national expert on state growth management and has consulted widely on the development of state policies. He is the author of numerous monographs including the pathbreaking *Land, Growth and Politics* (Chicago: American Planning Association, 1984).

Robert Einsweiler, AICP, is director of research for the Lincoln Institute for Land Policy in Cambridge, Massachusetts. He was formerly professor of Planning and Public Affairs at the Hubert Humphrey Institute of Public Affairs, University of Minnesota. He is the author of *Urban Growth Management Systems* (Chicago: American Planning Association, 1976) and *Strategic Planning* (Chicago: Planner's Press, 1988) and is co-author with John Bryson of *Shared Power* (Lanham, MD: University Press of America, 1991).

Michael Hibbard is associate professor in the Department of Planning, Public Policy, and Management and participating faculty in the Center on Human Development at the University of Oregon. He has published on social welfare policy and on the social impacts of economic change on resource communities.

Deborah Howe, AICP, is associate professor of Urban Studies and Planning at Portland State University. Her research interests include planning for an aging society and the commercial structure of cities. She has authored reports for the Department of Land Conservation and Development on developing a research program and growth management alternatives and has been on the board of directors of the Oregon chapter of the American Planning Association since 1986.

Gerrit Knaap is associate professor of Urban and Regional Planning at the University of Illnois. A specialist in urban and environmental economics, he has written extensively on health policy and on Oregon's growth management program. He is the co-author of *The Regulated Landscape: Lessons on State Land Use Planning from Oregon* (Cambridge, MA: Lincoln Institute of Land Policy, 1992).

Arthur C. Nelson, AICP, is professor of City Planning at the Georgia Institute of Technology. He has published extensively on growth management tools such as development impact fees and urban growth boundaries. He is the co-author of *The Regulated Landscape: Lessons on State Land Use Planning from Oregon* (Cambridge, MA: Lincoln Institute of Land Policy, 1992).

James R. Pease is professor of Resource Geography and Extension Land Resource Management Specialist, Department of Geosciences, Oregon State University. He has worked on rural land use research problems with state and local planners since 1973. His research and publications have focused on methods, techniques, and case studies of land use change, rating systems for resource lands, commercial agriculture and land use policy, environmental assessment, and the relationship of land tenure to resource degradation.

Mitch Rohse, AICP, is the communications manager for the Department of Land Conservation and Development. He is the author of DLCD's 1988 report on facility siting and *Land-Use Planning in Oregon* (Corvallis: Oregon State University Press, 1987).

Matthew I. Slavin is a research analyst with the Washington State Energy Office. He has previous experience in economic development and land use planning in Oregon and California and has recently completed a Ph.D. dissertation in Urban Studies at Portland State University dealing with the formulation and implementation of industrial policy in Oregon.

Edward J. Sullivan is a land use attorney with the firm of Preston Thorgrimson Shidler Gates and Ellis in Portland. He has been directly involved in many of Oregon's key land use cases, has authored essential law review articles on Oregon land use, and is recognized as a leading expert on the legal implementation of the Oregon system.

Nohad A. Toulan, AICP, is dean of the School of Urban and Public Affairs at Portland State University. His academic research and professional practice have focused on urban and regional planning in developing countries and on the evolution of American housing policy. In the mid-1980s he directed the preparation of a comprehensive plan for the holy city of Makkeh. More recently he has written and lectured about the importance of vision in the planning process.

Peter Watt is program consultant for the Lane Council of Governments in Eugene, Oregon. He is the author of the 1988 report *Policy Options for Siting Oregon Correctional Facilities* and "Siting Oregon's LULUs" in the Bureau of Governmental Research and Services's *Oregon Policy Choices, 1989.*

Index

A

Achterman, Gail 155
Acknowledgment xx, 4, 13–14, 40, 52, 53, 53–54; siting state facilities 149. *See also* public facilities: siting of
Adler, Sy 215, 218, 285, 292
Agricultural land goal 75
American Planning Association,: Oregon chapter 205
Arlington 151
Ashland 259
Association of Oregon Industries 76
Atiyeh, Vic xxi, 83, 84
AuCoin, Les 74

B

Baker xxi
Baker v. City of Milwaukie 51
Ballot Measure 5 xxii, 40, 41
Bassett, Edward 93
Bay City 259
Beale, Calvin 190
Beaton, C. Russell 30
Beaverton 110, 130
Bend xxi, 5, 34, 34–42, 41, 108, 191, 193
Benevolo, Leonardo 123
Benner, Richard 157
Benton County 183
Bernhardt, L. D. 176
Bettman, Alfred 93
Blake, Gerald 215
Blumenauer, Earl 74
Boal, Frederick 32
Bollens, Scott xxvii
Brookings 34, 35–42, 43, 108, 193
Bryant, C. R. 182
Buckingham, James S. 92
Burns, John 72, 73, 74, 75

C

California: land use planning 227
Calthorpe, Peter 213
Campbell, Larry 295
Capital improvement programs 27, 40
Carson, Everett 234
Center for Urban Studies 41
Cervero, Robert 140, 141
Challenges to Oregon program xx–xxii, 5, 211, 295
Chapin, F. Stuart 94
Citizen participation xxvi, 50, 56, 211, 221, 294
City of Happy Valley v. LCDC 104
City of Hillsboro v. Housing Devel. Corp. 103
Clackamas County 59, 108, 110, 151, 165
Clawson, Marion 30
Clinton, P, J. 170
Cogan, Arnold 210
Colorado: land use planning 227, 259
Comprehensive plans 50, 51, 130
Concurrency xxvii, 141, 227, 231, 236, 237, 238, 239, 241
County Planning Commission Act (HB 2548) 164
Curry County 247
Curry County v. 1000 Friends of Oregon 165
Cusma, Rena 44

D

Daniels, T. L. 175, 176, 177
Daughton, K. 182
Davidoff, Paul 97, 98
Day, L. B. xviii, 78
Deakin, Elizabeth 141
DeGrove, John 72, 80, 213, 228, 232, 291
Densities, residential 106–120; maximum 39; minimum 18, 36, 39, 106, 108, 293
Department of Environmental Quality (DEQ) 151
Department of Land Conservation and Development (DLCD) xix, 8, 49, 121, 127–131, 131–139, 143, 157, 158, 206, 275; and citizen participation 210; conflict with local governments 27; consistency 18; monitoring and evaluation 276, 277. *See also* Evaluation of Oregon program; research 280
Deschutes County 108
Dickas v. City of Beaverton 11
Dodds, Gordon 209

323

Doughton v. Douglas County 63–64
Douglas County 184, 190, 192, 193
Drain 192
Dyckman, John 97

E

ECO Northwest 34, 36, 38, 41, 109, 115
Economic development 79–81, 83–85; and growth management 71–75, 75–79, 79–81; and the Oregon program xxi, 71–87; planning 198, 199
Economic Development Commission (EDC) 73, 77, 85
Economic Development Department (EDD) 76, 78–79, 79–81, 81–83, 198
Economic development goal 75, 195
Edner, Sheldon 215, 218
Einsweiler, Robert 284, 288
Elazar, Daniel 206, 208
Eugene xvii, xxi, 5
Evaluation of Oregon program 275–279, 291–293; research agenda 279–290
Exception areas xxiv, 33, 43, 165, 246, 260, 269, 286
Exclusive farm use (EFU) zones 164, 169, 266, 284

F

Fadeley, Nancie xviii
Farm land 5, 14, 25, 32–33, 50, 147, 167–175, 228; large-scale vs. small-scale 167–175, 170–185, 175–183, 183–184. *See also* Secondary lands
Farm Land Tax Assessment Act (SB 101) 164
Farming operations 32
Fasano v. Board of Commissioners of Washington County 12, 51, 63
Ferguson, Erik 142
Fischel, William 32, 180
Florida: land use planning xxvii, 33, 141, 163, 227, 228–242, 238, 239, 258
Forest Grove 110
Forest land 14, 50, 167–175, 228
Fourier, Charles 92

G

Gans, Herbert 97
Garreau, Joel 248
Gastil, Raymond 207
Geddes, Patrick 95
George, Henry 264, 265, 268
Glendale 193
Goal 10 99, 100–102, 102, 104, 106, 107, 112, 211, 292
Goal 11 149, 211, 246
Goal 12 121, 149
Goal 13 211
Goal 14 35, 75, 99, 100–102, 102, 106, 107, 108, 196, 211
Goal 3 75, 171, 211
Goal 4 172
Goal 5 211
Goal 8 99, 195
Goal 9 71, 75, 80, 81–83, 195
Goals. *See* State planning goals and individual goals by name and number.
Gold Beach 6
Goldschmidt, Neil 84, 213
Goldschmidt, W. 168
Goldy, Dan 78
Grants Pass 43, 191, 193
Gresham 112
Growth management 71–87, 75, 75–79, 83–85, 98, 99–113, 113–116
Gunther, John 206

H

Hallock, Ted xviii
Halprin, Lawrence xviii
Hansell, Stafford xxi
Happy Valley xxv, 112
Hart, John 261
Hawaii: land use planning 163, 227
Hibbard, Michael 72, 283
Hood River 43
Hood River County 173
House Bill 2225 153
House Bill 2548 164
House Bill 2713 152, 156
House Bill 2884 155
House Bill 2936 155
House Bill 3092 151, 155
House Bill 3661 174, 221, 295, 296

Index

Housing 31, 50, 55, 91–120, 285; affordable xxv, 29, 37, 45, 102, 113–116, 140, 228, 292; and growth management 99–113; Comprehensive Housing Affordability Strategy 112–117; "fair share" 102–105; history of regulation 92–99; "least cost" 102–105; low-income 101, 113–116; prices 31. *See also* Urban growth boundaries: and land values; role of federal government 94–96; role of the states 96–100; St. Helens Policy 102–105
Howard, Ebenezer 95
Howe, Deborah 284, 288

I

Implementation 14–16
Innes, Judith xxvii
Integrating land use and transportation 292. *See also* Transportation

J

Jackson County 173
Johnson, Lyndon 95
Joint Legislative Committee on Land Use 10, 52, 166, 169, 180; and secondary lands 171
Joint Policy Advisory Committee on Transportation 125, 206, 217

K

Kent, T. J. 93
Klamath Falls xxi
Klingman, David 208
Knaap, Gerrit 30, 31, 36, 38, 42, 269

L

La Grande 104
Lake Oswego 104
Lammers, William 208
Land Conservation and Development Commission (LCDC) xiv, xv, 9, 10, 49, 71, 75–76, 78–79, 79–81, 81–87, 121, 122, 156, 158, 164, 276; acknowledgment 8; activities of statewide significance 153, 154; and housing goal 101; and housing needs assessment 104; and secondary lands 172, 220; and statewide planning goals xix; and urban growth boundaries 27; and urban reserve rule 43; and urbanization goal 26; state agency coordination 19; study of the effectiveness of farm and forest policy 180, 181, 182, 183
Land Use Board of Appeals (LUBA) xv, xx, 10, 18, 49, 60–63, 122, 165
Land Use Planning Act. *See* Senate Bill 100
Land values 252–259. *See also* Urban growth boundaries: impact on land values; government intervention 252–259
Lane County 192
Legislature, Oregon 8, 10–11, 71–75, 79, 83, 85, 160, 295
Leonard, Jeffrey 191
Liberty, Robert 232
Linn County xviii
Little, Charles 245, 246
Local governments, role of 11–12, 13–14, 14–16, 18, 49, 56, 59–64, 63–66, 127, 133, 147, 148–150, 163, 294

M

Macpherson, Hector xviii, xix
Maine: land use planning 233–242, 291
Marginal lands. *See* Secondary lands
Marion County 25, 151
Marston, Ed xxiv
Mayer, Tanya 261
McCall, Tom xvii, xviii, 5, 50, 75, 76, 96, 208
McDonough, M. 182
McHarg, Ian 97, 259
Medford xxi, 5, 34, 35–42, 43, 108
Metropolitan Homebuilders Association 18, 106; monitoring and evaluation 278
Metropolitan Housing Rule 105, 105–110, 111–113, 114–116, 139, 293; and affordable housing 114–116
Metropolitan Service District 44, 53, 59, 105, 111, 112, 121, 126, 132, 137, 139, 206, 212, 213, 218, 231

Michaud, Mike 236
Milwaukie xxv
Ministerial decisions 63–66. *See also* Quasi-judicial decisions
Moralistic political culture 207–210
Mosser, John 79
Multnomah County xix, 59, 110, 112, 151
Muth, Richard 32

N

National Growth Management Leadership Project 231, 234, 243
Nelson, Arthur 30, 31, 32, 33, 34, 36, 38, 175, 176, 177, 178, 269
New Jersey: land use planning xxvii, 141, 231–242, 257–258, 258, 261
Newberg 43
Newport 104
Nixon, Richard 96
North Carolina: land use planning 227

O

Oakridge 192
Ocean Resources Management Plan 293
1000 Friends of Oregon xv, xxvi, 18, 50, 109, 132, 143, 163, 164; 1982 housing study 109; 1991 housing study 106, 114; and growth management in Florida 228, 229, 230, 231; and growth management in Maine 234; and growth management in New Jersey 232; and growth management in Washington state 238; and housing densities 109; and interest group research 277, 278; and Metropolitan Housing Rule 110; and National Growth Management Leadership Project 243; and secondary lands 172; and statewide transportation goal 129; and transportation planning rule 132, 133, 134, 135, 136. *See also* transportation rule; and urban form objectives 130; legal challenges by 129; opposition to Rajneeshpuram 216; opposition to western bypass 122, 126; studies of county permitting procedures 168, 169, 183; supports LCDC and DLCD 13; watchdog role of 10, 19, 127, 291

1000 Friends of Oregon v. Lane County and LCDC 172
Open space preservation 140
Oregon City 110
Oregon Department of Transportation (ODOT) 122, 126, 127–131, 131–139, 143, 199
Oregon Progress Board 279
Owen, Robert 92

P

Pease, James 284
Peirce, Neal 206, 209, 238
Pendleton xxi
Periodic plan revisions 27
Periodic review xx, 52, 54, 56, 106. *See also* Post-acknowledgment review; Review process
Permit process 64–67
Peter Wilson and Associates 267
Peterson v. City of Klamath Falls 11
Polk County 25, 151, 165
Popper, Frank xxiv, xxvii
Population 33; urban vs. rural 190–191
Population growth 126
Portland xvi, xvii, xxi, xxii, xxiii, xxvi, 5, 18, 27, 33, 34, 34–42, 43, 55, 59, 104, 108, 109–113, 125, 130, 135, 139, 206, 214–216, 219, 261, 267, 293; Downtown Plan 111, 206, 215–216, 217; Office of Neighborhood Associations 206, 214–215
Post-acknowledgment review 55–56. *See also* Periodic review
Prineville 6
Property rights 29
Public facilities 27, 31, 39, 39–45, 257–258; siting of 148–162
Public facilities and services goal 27, 76

Q

Quasi-judicial decision making 50
Quasi-judicial decisions 60, 63, 63–66. *See also* Ministerial decisions
"Quiet revolution" 7

Index

R

Rajneeshpuram 206, 216–217, 285
Ramis, Tim 42
Reagan, Ronald 96
Recreational needs goal 195
Redmond xxi, 104
Regional Strategies program 84–85
Regional Urban Growth Goals and Objectives (RUGGO) 212–213
Reiner, Thomas 97
Review process 4. *See also* Periodic review; Post-acknowledgment review
Rhode Island: land use planning 241–242, 291
Richmond, Henry 228, 231, 232
Riddle 192
Robbins, William xxiii
Roberts, Barbara 157, 174, 295
Rohse, Mitch 168
Roseburg 6
Ross, James 184
Rule making 54–55, 121, 121–122, 125–127, 127–131, 131–139
Rural communities 284; economic development 190–201
Rural development xxiv, 164, 245–273
Rural land xvi, 163–188, 165, 189–201, 284
Russwurm, L. H. 182

S

Salem xxi, 5, 6, 25, 27
Sandy 43
Seaman et al. v. City of Durham 11, 103
Secondary lands xxiii, 170–185, 220, 221, 294. *See also* Farm land: large-scale vs. small-scale
Seltzer, Ethan 44
Senate Bill 10 99; compare with Washington state law 291; precursor to Senate Bill 100 xvii; precursor to Senate Bill, 100 xvii
Senate Bill 100 xiv, xvi, 3, 72, 79–81, 99, 100, 147, 148, 164; acknowledgment 53, 75; amendments to 10; and economic development 74, 76; and housing policy 111; citizen participation 210; compare with Washington state law 291; compliance with mandates 81; contrast with Senate Bill 224 78, 79; contrast with Washington state law 237, 240; legislative history xviii, xix, 8, 8–9, 9, 82, 152; opposition to 73; Vic Atiyeh supports 82, 83
Senate Bill 101 164
Senate Bill 224 72, 72–75, 79–81, 81–83, 85
Senate Bill 389 155
Sharkansky, Ira 207, 208
Sinclair, Robert 32
State agencies, role of 19, 57–59, 127–146
State agency coordination 57, 59, 128, 134; siting facilities 150. *See also* Public facilities: siting of
State of Oregon v. City of Forest Grove 103
State planning goals xix, 9, 49, 71, 99, 149, 164, 211, 287–289; adoption or amendment 52; enforcement 55; referendum challenges 211. *See also* Challenges to Oregon program
Straub, Bob 77, 78, 82
Sullivan, Ed 102, 284
"Supersiting" laws. *See* public facilities: siting of
Support for Oregon program 6–7
Sustainable development 297

T

Thompson, Jr., E. 168
Toulan, Nohad 285
Tourism xxii
Transportation 121–146; alternatives to automobile 137, 139; history of planning 123
Transportation goal 76, 121
Transportation rule xxiii, 121–146, 292
Tri-Met 126

U

Umatilla County xxi
Unemployment 83, 192
Urban growth boundaries xx, xxii, xxv, 11, 25–47, 50, 196, 197, 206, 228, 245, 266, 286; after the year 2000 42–44; and control of urban sprawl 292; and farm land 32–33; and growth management in Washington state 238, 239; and growth rates 267; and

highway planning 126, 132; and housing 31–32; and housing planning 293; and housing prices 113, 114; and land values 29–30, 30, 246, 250, 268; and nonresource residents 182; and public facilities 39; and small town development 196, 197; and speculation 260; and supply of buildable land 106; and western bypass 121; comparison with Florida 229; contrast with Washington state 240; density of development 16, 108; development outside of 108; extensiveness of 12, 13; impact on land values 100, 246; in small towns 196–197; in the Portland metropolitan area 212; interest in Maine 234; model for New Jersey 233; residential development inside and out 34–42; urban and urbanizable land within 27
Urban reserves 43, 45
Urbanization goal 26, 35, 38, 75

V

Vermont: land use planning 163, 227, 241, 242, 268
Vogel, David xv

W

Wachs, Martin 142
Waldo, Dwight 214
Wasco County 217
Washington County 59, 112, 121, 125, 129, 130, 132, 151
Washington state: land use planning 237–242, 291
Weber, Max 214
White, Charles 215
Whitelaw, W. Ed 30, 78
Whyte, William 97
Willamette River Greenway xvii, xix, 52, 55, 209
Willamette Valley xvii, 5, 6, 7, 9, 72, 77, 84, 169, 173, 194, 284, 292

Z

Zoning 13, 14–15, 51, 255–256